Wastewater Engineering

and Treatment 1

Wastewater Engineering and Treatment 1

Composition, Pollutant Loads, Primary Settling, Activated Sludge

Kader Gaid

ISTE

WILEY

First published 2026 in Great Britain and the United States by ISTE Ltd and John Wiley & Sons, Inc.

ISTE Ltd
27-37 St George's Road
London SW19 4EU
UK

www.iste.co.uk

John Wiley & Sons, Inc.
111 River Street
Hoboken, NJ 07030
USA

www.wiley.com

Library of Congress Control Number: 2026933319

British Library Cataloguing-in-Publication Data
A CIP record for this book is available from the British Library
ISBN 978-1-83669-055-9

The manufacturer's authorized representative according to the EU General Product Safety Regulation is Wiley-VCH GmbH, Boschstr. 12, 69469 Weinheim, Germany, e-mail: Product_Safety@wiley.com.

Contents

Introduction ix

Chapter 1. Urban Wastewater Composition 1

1.1. Main physicochemical parameters of raw wastewater 1
1.1.1. Suspended solids 3
1.1.2. Organic matter indicators 5
1.1.3. Nitrogen 15
1.1.4. Phosphorus 19
1.1.5. Micropollutants 20
1.1.6. Parametric influence in the treatment system. 24
1.2. Microbiological parameters: pathogenic organisms and indicators of fecal contamination 25
1.2.1. Bacteria 29
1.2.2. Enteric viruses. 31
1.2.3. Enteric protozoa. 32
1.2.4. Helminth eggs. 32
1.2.5. Fecal contamination indicators: assessment in treated wastewater 34
1.3. References 38

Chapter 2. Processing Unit Operations 43

2.1. Background. 43
2.2. Design principles for treatment plants 45
2.3. Unit operations. 48
2.3.1. Pretreatment 51
2.3.2. Primary treatment. 52
2.3.3. Secondary treatment 52
2.3.4. Effluent clarification and discharge 54

2.3.5. Additional treatments 54
2.3.6. Sludge treatment 58
2.4. References 62

Chapter 3. Estimation of Pollutant Loadings 65

3.1. Water consumption 65
3.2. Estimated discharges 66
3.2.1. Wastewater flows 67
3.2.2. Peak flows 68
3.2.3. Number of population equivalents (PE) 69
3.2.4. Maximum and minimum flow rates 71
3.3. Pollutant loads 74
3.3.1. Relationship between load and concentration 74
3.3.2. Pollutant loads in BOD, N and P 75
3.4. References 76

Chapter 4. Pretreatments 79

4.1. Screening 79
4.1.1. Manual screens 80
4.1.2. Mechanical screens 81
4.1.3. Screen calculation 83
4.1.4. Screen implementation 89
4.2. Comminution 89
4.3. Sieving 90
4.3.1. Pressure washing sieves 90
4.3.2. Mechanical waste discharge sieves 90
4.4. Sand removal 91
4.4.1. Calculation for aerated sand trap 100
4.5. Oil and grease removal 100
4.5.1. Grease separators 101
4.5.2. Aerated grease and oil trap 101
4.6. References 105

Chapter 5. Primary Settling 107

5.1. General principles 107
5.2. Conical or cylindrical–conical static settler 109
5.3. Horizontal flow settler 110
5.3.1. Dimensioning of horizontal settlers 111
5.4. Circular settlers 117
5.4.1. Example design of a primary settler 120
5.5. Design of primary mechanized settlers 123
5.5.1. Rectangular settler 123

5.5.2. Circular settler. 124
5.6. Performance of conventional primary settlers. 125
5.6.1. Settler depth 128
5.7. Lamellar settlers 129
5.7.1. Theory and general principles 130
5.7.2. Dimensioning basics for lamellar settlers. 133
5.8. References 174

Chapter 6. Biological Treatment with Activated Sludge 179

6.1. Background information on microbial organisms. 179
6.2. Operating principle 181
6.3. Theoretical bases of aerobic treatment. 182
6.3.1. Organic matter (biological oxygen demand) evolution in the presence of bacterial mass 182
6.3.2. Analytical interpretation 184
6.4. Traditional design methods 192
6.4.1. Complete mix 194
6.4.2. Plug flow 211
6.4.3. Comparison between the two systems 213
6.4.4. Oxygen requirements for carbon pollution 213
6.5. Operating parameters 221
6.5.1. Sludge losses: causes and consequences 221
6.5.2. Role of nutrients 223
6.5.3. MI 223
6.5.4. DI. 224
6.5.5. Sludge redox potential (Rh) 224
6.5.6. Effect of temperature 225
6.5.7. Effect of pH 226
6.5.8. Effect of some toxicants 226
6.5.9. Sludge settling issues 227
6.6. Activated sludge ecology 230
6.7. Aeration of activated sludge tanks 233
6.7.1. O_2 transfer 233
6.7.2. Influence of some factors on Cs and K_La 236
6.7.3. Aeration systems 243
6.7.4. Surface aerators 249
6.7.5. Deep-water aerators. 252
6.7.6. Conditions of use of aeration systems. 253
6.7.7. Aerator design. 257
6.7.8. Using pure oxygen 277

6.8. Activated sludge processing technology 278
6.8.1. Conventional activated sludge 278
6.8.2. Contact stabilization 284
6.8.3. Full mixing 286
6.8.4. Sequencing batch reactor 287
6.8.5. Bioring™ process 289
6.9. Examples of dimensioning the activated sludge process 291
6.9.1. Example 1: calculation of a low load aeration tank 291
6.9.2. Example 2: calculation of an aeration tank fed under medium loading . . 293
6.9.3. Example 3: population of 3,000,000 inhabitants 295
6.9.4. Example 4: population of 12,500 inhabitants 302
6.10. References 307

Index 313

Introduction

At a time when demographic and industrial growth conditions our environment and our way of life, it is now more important than ever that we focus on everything that can harm our well-being and living conditions.

It is true that industrial deployment and expansion result in huge benefits at the economic level and bring about significant improvements in terms of social standards. Nonetheless, the major drawback is that this industrial boom incurs biological, physical and chemical hazards (pesticides, food additives, hydrocarbons, etc.), which in the short or long term affect us through the food chain on which we depend.

In parallel, the working population that left the countryside has moved closer to and developed around industrial centers, thus creating large cities, whose inhabitants discharge their polluting flows into a receiving environment such as a lake, a river or a sea on a daily basis.

Industrial (IWW), municipal (MWW) and domestic (DWW) wastewater contains non-negligible amounts of organic, inorganic and microbial pollutants. This wastewater is one of the main sources of contamination of surface- and groundwaters. Effective treatment strategies will have to be adopted in order to optimally achieve discharge standards, which tend to become increasingly more severe, both in terms of pollutant concentrations and flows. For example, for DWW discharges into the natural environment, regions referred to as normal areas have been defined, for which the treatment primarily relies on the removal of particulate fractions and carbon pollution, while in regions known as sensitive areas, there are nitrogen, phosphate and microbial fractions for which additional removal is required. For EWWs, said conventional processes, be they intensive (activated sludge, trickling filters, aerobic biofilters, etc.), extensive (lagoon, infiltration system, etc.) or combined processes, can address discharge requirements by offering more or less reliable performance, due to their sensitivity to sudden variations in the loads to be treated (for suspended culture systems), the flocculation state of the biological populations (for activated sludge) or failure in controlling biomass and flow distribution within the porous beds (attached culture systems in self-cleaning biofilters, or biofilters). For IWWs, regulations also specify the conditions for discharge into the natural environment or the municipal sewer system.

Among all the possible water treatment solutions, there is one that uses the natural benefits offered by nature, by means of microorganisms comprised in the ecological system. These are processes for the development and control of microorganisms, the main objective of which is the removal of organic pollutants by transforming them into mineral salts.

Wastewater Engineering and Treatment 1 presents the theory and practice of wastewater treatment in an educational and balanced way, so that the selection, design and operation of wastewater treatment plants can be performed. The theory is considered essential to understand the principle of wastewater treatment operation. Practice is associated with the direct application of concepts and modes of usage. The main wastewater treatment processes are outlined, including the characterization of wastewater and the design of the wastewater treatment system up to the choice of sludge treatment and disposal units.

This book provides a broad overview of the biological treatment of wastewater and the treatment and recycling of sludge. Most of this work consists of unit operations and the various processes and industries associated with biological wastewater treatment. The main topics covered are: wastewater treatment microbiology and ecology; reaction kinetics and reactor hydraulics; the transformation and degradation of organic and inorganic matter; settling and clarification as well as aeration.

This book can be used for the purpose of engineering courses in environmental engineering, environmental science and related fields, and also in the context of disciplines related to wastewater treatment.

The succession of chapters was devised such as to follow the path of the water through the treatment plant and they are structured according to the main wastewater treatment processes (activated sludge process, trickling filters, biofilters, lagoons, biological discs, membranes, hybrid systems with mobile supports).

The main technologies and their variants are comprehensively covered, including the main concepts, operating principles, scaling and implementation.

The last chapter covers sludge treatment and recycling. It deals with the characteristics of sludge; unit operations such as thickening, digestion, drying, stabilization and the reduction of pathogen microorganisms; and recovery through spreading, composting and the formation of by-products such as struvite or bioplastics.

1

Urban Wastewater Composition

Domestic wastewater contains about 99.9% water. The residual part is comprised of dissolved suspended organic and inorganic matter, as well as microorganisms. These 0.1% are what constitute water pollution and require that it be treated.

1.1. Main physicochemical parameters of raw wastewater

When designing a wastewater treatment plant (WWTP), it is useful to determine the different compounds present in the wastewater. Treated wastewater contains solids, dissolved substances and microorganisms. These latter are the main cause of restrictions imposed on reusing wastewater because they very often contain elements that are harmful to health, such as organic micropollutants and pathogen agents such as viruses, bacteria, protozoa and helminths. The load in pathogens in raw water is essentially a function of the health status of the population from which the wastewater originates. The risk of waterborne infection by these pathogens depends on a combination of a number of factors that include their concentrations, their dispersion in water, the capacity of these intestinal agents to survive in the environment, the quality of the water purification treatment, the infective dose, the exposure and the susceptibility of the exposed population.

The main prevalent parameters in treated wastewater that deserve special attention are:

– the indicators of organic matter (biological oxygen demand and chemical oxygen demand);

– suspended solids;

– mineral substances;

– heavy metals;

– microbiology.

Carbonaceous organic matter existing in urban wastewater is made up of a heterogeneous mixture of organic compounds of various structures and variable origin, including natural organic matter, soluble microbial products, residues of pharmaceutical products, protein compounds, carbohydrates, lipids, urea, surfactants, phenols, industrial residues, bacterial degradation by-products, pesticides and endocrine disruptors.

This great heterogeneity in the composition of urban wastewater must be taken into account when reusing it to be consumed as direct or indirect drinking water or for specific industrial applications, groundwater recharge, or irrigation of food eaten raw.

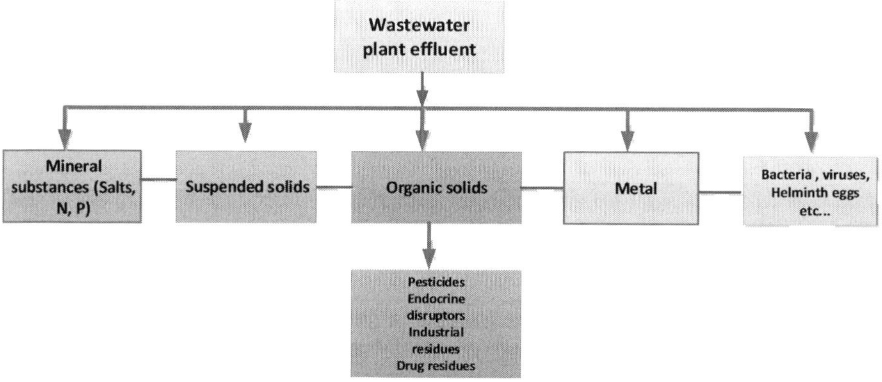

Figure 1.1. *Physicochemical and microbiological characteristics of wastewater*

The microbiological composition of wastewater is extremely variable according to the season, the origin of the collected wastewater, the population health status, etc. As a result, it contains a wide variety of microorganisms, with varying concentrations, potentially pathogenic to humans (bacteria, molds and their toxins, viruses and parasites) and likely to induce health hazards via respiratory, mucocutaneous and/or ingestion routes depending on how sensitive the person exposed is and the dose of microorganisms a person is exposed to (Michael-Kordatou 2015).

1.1.1. *Suspended solids*

Suspended solids (SS, or suspended particles) refer to insoluble solids that can be seen with the naked eye present in suspension in a liquid. Total suspended solids (TSS) is a measure of suspended solids.

Suspended solids are also studied as a "matrix" capable of adsorbing various pollutants, which can be transported along with the water in treatment plants. A distinction is made between:

– colloidal suspended solids (between 10 nm and 1 μm in size);

– dissolved solids (less than 10 nm in size).

The fine suspended particles in treated wastewater either have a natural origin in connection with precipitation or an anthropogenic origin and are then brought by atmospheric deposition and/or by urban, agricultural, domestic, rural and industrial discharges. SS can be classified according to their size and condition, chemical characteristics and settleability.

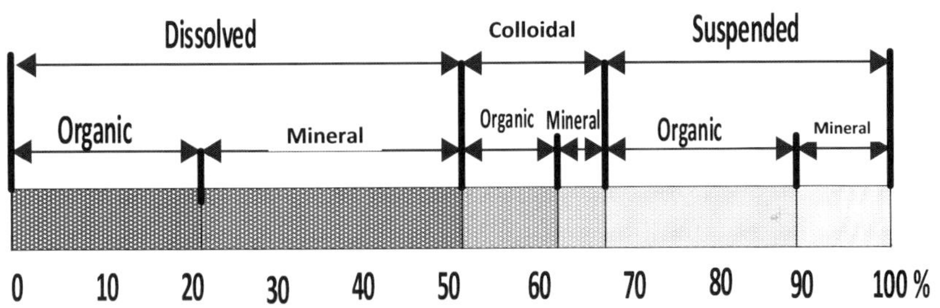

Figure 1.2. *Typical distribution of suspended solids in urban wastewater*

Raw municipal wastewater, which is a mixture of domestic wastewater and industrial wastewater, varies in composition by the hour, by the day and from one place to another. The total solid content (TSC) of raw wastewater (i.e. the sum of species of dissolved colloidal and suspended solids) depends on water consumption per capita and the nature of industrial discharges into the sewer drainage basin. The TSC value usually ranges between 500 and 1,000 $mg.L^{-1}$. A typical fractional characterization of solid matter in raw municipal wastewater into organic, inorganic, dissolved and suspended species is presented in Figure 1.2.

The inorganic particulate fraction of municipal wastewater presents a much higher density and sedimentation rate than the organic particulate fraction. It is partly eliminated during the first stage of treatment. The remaining particle size ranges from 10 mm to colloidal size. Approximately 55–65% of total suspended solids (TSS) in municipal wastewater are settleable and therefore would be removed by a gravity sedimentation process.

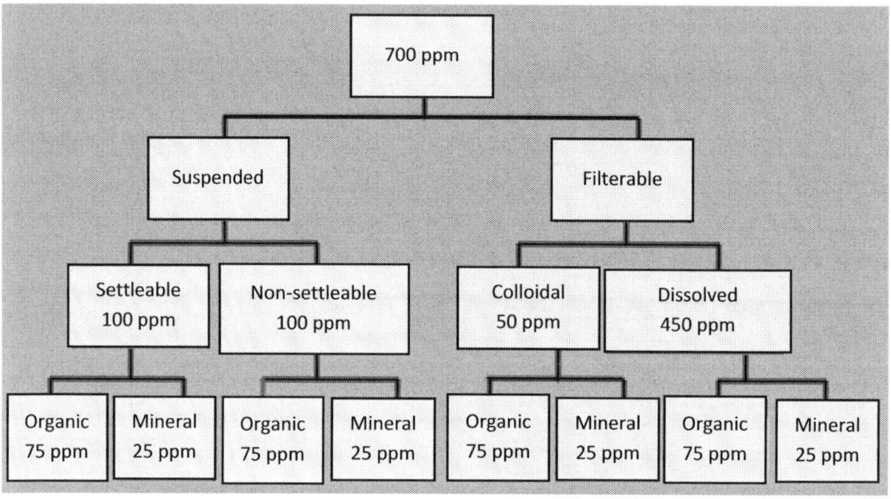

Figure 1.3. *Example of suspended solid distribution*

In general, dissolved solids are those with a diameter smaller than 10^{-3} μm, colloidal solids are those with diameters comprised between 10^{-3} and 100 μm and suspended solids are those with a diameter greater than 100 μm.

The term "particulate" is used to indicate that matter existing in wastewater is present in the form of particles. These are called suspended solids. In this context, expressions such as particulate BOD, particulate COD and particulate phosphorus serve to indicate that the parameters are related to suspended solids. In contrast, soluble BOD, COD and phosphorus are associated with dissolved substances.

Weight materials present in the effluent are subdivided into various forms which can be represented by:

– suspended solids (SS);

– volatile suspended solids (VSS);

– suspended mineral solids;

– settleable solids;

– non-settleable solids.

Suspended solids are neither solubilized nor colloidal. It may be considered that they represent an intermediate between mineral sand or coal dust particles and mineral particles of the mucilaginous type. They contain organic matter and mineral matter. Two techniques are currently used for dosing suspended solids:

– filtration separation (paper filters, membranes);

– centrifugation.

Suspended solid contents are obtained after drying at 105°C a known sample volume. SS concentrations in wastewater are highly variable, ranging from about 250 to 400 $mg.L^{-1}$. Volatile suspended solids represent the organic fraction of suspended solids. They are measured by calcination at 550°C of a sample whose SS content is already known. They make up about 70–80% of SS. When solids are submitted to a high temperature (550°C), the organic fraction is oxidized (volatilized), leaving only the inert (non-oxidized) fraction after combustion. Volatile solids represent an estimate of organic matter in solids, while non-volatile (fixed) solids represent inorganic or mineral matter.

Mineral solids represent the difference between suspended solids and volatile solids. They indicate the weight of calcination residues and correspond to the presence of salts, silica or dust, for example. Settleable solids are considered to be capable of settling during a one-hour period. The volume of solids accumulated at the bottom of a container called an Imhoff Cone is measured and expressed in $mL.L^{-1}$. The fraction that does not settle represents non-settleable solids. Suspended solids (SS) are mostly biodegradable in nature. The largest proportion of pathogen microorganisms contained in wastewater are transported by SS. In treated water, the limit concentration of suspended solids of 35 $mg.L^{-1}$ requires much more advanced treatment to deal with health risks and applied technologies. For example, excessive suspended solids can lead to complications in the transport and distribution of effluents as well as clogging of irrigation systems.

1.1.2. *Organic matter indicators*

The organic matter present in wastewater is an important parameter that is the source of the main problems of water pollution. The organic substances present (Pessoa and Jordao 1982) in wastewater are mainly composed of:

– protein compounds (≈ 40%);

– carbohydrates (≈ 25 to ≈ 50%);

– oils and fats (≈ 10%);

– urea, surfactants, phenols, pesticides and others (< 10%).

Carbonaceous organic matter (organic carbon-based) present in wastewater can be divided into several fractions:

– suspended (particles);

– dissolved (soluble);

– inert;

– biodegradable.

Direct or indirect methods are used for the quantification of organic matter:

– indirect method: measurement of oxygen consumption:

  - biochemical oxygen demand (BOD),

  - ultimate biochemical oxygen demand (uBOD),

  - chemical oxygen demand (COD);

– direct method: measurement of total organic carbon (TOC).

1.1.2.1. *Biological oxygen demand (BOD_5)*

The main ecological effect of organic pollution in a body of water is the decrease in the concentration of dissolved oxygen. The wastewater pollution potential is evaluated by measuring oxygen consumption. This quantification can be obtained by stoichiometric calculations of the oxidation reactions of organic matter. If the substrate is, for example, glucose ($C_6H_{12}O_6$), the amount of oxygen required to oxidize this substance can be easily calculated. This is the principle of theoretical oxygen demand (TOD).

However, wastewater exhibits great heterogeneity in its compositions, and it is not possible to establish a list of all its constituents and to calculate the oxygen demand according to the chemical oxidation reactions of each of them. The solution is to measure in the laboratory the oxygen consumption by a standard volume of wastewater in a given time interval. The concept of biological oxygen demand (BOD) was thus introduced. BOD represents the amount of oxygen needed to oxidize organic (biodegradable) matter by biological means (oxidation of biodegradable organic matter by bacteria). Biological oxygen demand (BOD) is a water pollution index used for evaluating the biodegradable fraction of the carbon pollutant load in wastewater, and is generally calculated after five days at 20°C and protected from light: this is referred to as BOD_5.

BOD is an evolutionary phenomenon in the sense that it enables the study of the behavior of an organic load in the presence of microorganisms. BOD can only be determined if the existing microorganisms are able to simulate the organic substances present in the water.

The evolution of a sample of urban wastewater reveals a representation of organic matter that is always identical, in two successive stages (see Figure 1.2):

– The first stage relates to carbon compounds. It initiates quickly and is completed in 20 days, working at a temperature of 20°C.

– The second stage corresponds to the oxidation of nitrogenous compounds, which only initiates after the tenth day, due to the slow kinetics of the bacteria that convert ammoniacal nitrogen into nitrites and nitrates.

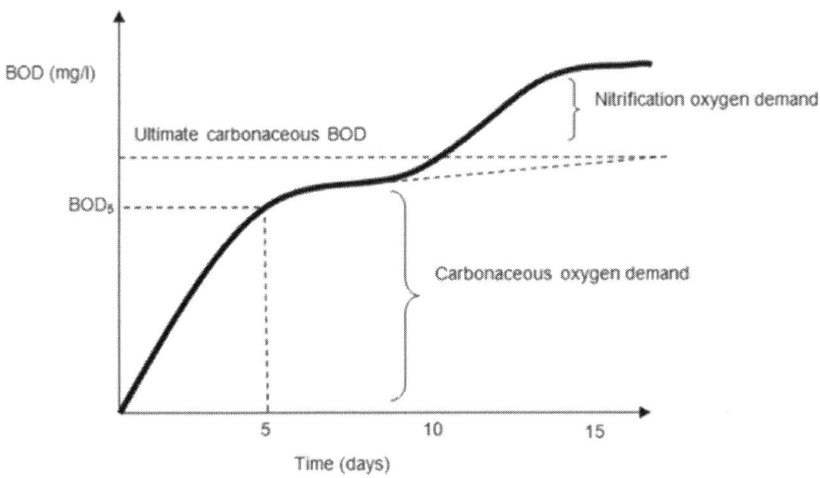

Figure 1.4. *BOD evolution at 20°C*

In fact, studies have resulted in a mathematical law of BOD evolution:

$$BODu = BOD\infty\ (1\text{-}e^{-kt})$$

where the BOD∞ is the BOD corresponding to the total mineralization of carbonaceous organic matter and k is a variable factor of about 0.2 for urban wastewater.

Therefore, at 20°C, on the first day, 20.6% of the BOD of the first stage disappears, on the second day 20.6% of the remaining BOD disappears, that is, 20.6% of the remaining 79.4%. On the third day, 20.6% of this remaining fraction disappears, etc., as shown in Table 1.1.

Oxidation time (days)	**1**	**2**	**3**	**4**	**5**	**6**	**7**	**8**
% oxidation (1st stage) at 20°C	20.6	37	50	60	68.4	75	80	84
Oxidation time (days)	**9**	**10**	**12**	**14**	**16**	**18**	**20**	**25**
% oxidation (1st–2nd stage) at 20°C	87	90	94	96	97.5	98.4	99	99.7

Table 1.1. *O_2 consumption over time in a wastewater sample*

In practice, BOD measurements are limited to oxygen demand during the first five days: this is the BOD_5 measured at 20°C. It should be noted that the daily fraction of consumed BOD is different depending on the temperature, as shown in Table 1.2.

Temperature °C	5	10	15	20	25	30
% consumed	10.9	13.5	16.7	20.6	25.2	30.5

Table 1.2. *Consumed BOD according to incubation temperature*

In practice, complete stabilization takes several days (about 20 days or more for domestic wastewater). This corresponds to the ultimate biological oxygen demand (uBOD). However, to shorten the duration of the laboratory test and obtain results in a reasonable time, standardization was established.

BOD is determined after five days, hence the term BOD_5. The test is carried out at a temperature of 20°C. Ultimate BOD is expressed in BOD_{20}.

The BOD test follows a fairly simple protocol. BOD is commonly determined using the dilution method. It consists of diluting the water to be analyzed in a certain ratio, with clean water saturated with dissolved oxygen. The first step consists of determining the quantity of dissolved oxygen remaining after this mixture, because there is an immediate demand from the first minutes, caused by mineral or organic reducing substances.

A second fraction of this mixture is stored at 20°C for five days, after which the amount of residual oxygen is determined.

The product of the difference ($O_{22}(0)$ – $O_2(t)$) multiplied by the dilution ratio gives BOD_5.

Dilution water is the subject of much controversy. Some authors recommend the use of rural river water, as long as this water is not itself polluted. When no clean river water is available, an attempt is made to reproduce it from distilled water by addition of nutrient salts such as phosphates (sodium and potassium) and chlorides (calcium and magnesium; ammonium and ferric). These results allow the obtaining of synthetic dilution water. This water is next aerated to saturation and then used for analysis.

When the water to be analyzed is low in organic substances, the dilution water is inoculated with pure bacterial strains in a defined quantity, or by adding a fraction of river water. Some even suggest not washing the incubation vials so that there is self-seeding. This may be the case when it is always the same type of water that is analyzed (continuous controls of influents and effluents at the treatment plant).

It is recommended that different dilutions be done with several samples to obtain the true BOD value. The analysis is actually subject to variations due to pH and the presence of toxic substances (metal ions, organic products) that inhibit bacterial activity in the measurement vial and reductants (such as mercaptans, for example).

For wastewater, some practical aspects require a few adaptations. The high concentration of organic matter quickly consumes all the dissolved oxygen and prevents any analysis. Dilutions are then applied in order to reduce the concentration of the organic matter, such that the oxygen consumption on the 5th day is lower than the oxygen initially present in the sample. Domestic wastewater has a BOD_5 of approximately 300 $mg.L^{-1}$, meaning that 1 liter of wastewater is associated with a consumption of about 300 mg of oxygen, in five days.

The main benefits of BOD testing are:

– an approximate indication of the biodegradable fraction of the wastewater;

– an indication of the pollution level in the wastewater;

– an indication of the rate of oxygen consumption with respect to time;

– an approximate determination of the amount of oxygen required to biochemically stabilize the organic matter present.

The following limitations have, however, been mentioned (Marais and Ekama 1976):

– Low BOD_5 levels can be observed when microorganisms responsible for decomposition are not adapted to the existing organic substances.

– Heavy metals and other toxic substances can inhibit microorganisms.

– The uBOD/BOD_5 ratio varies according to wastewater.

Despite these limitations, the BOD test is still widely used, especially if we bear in mind that the design criteria for many wastewater treatment processes are expressed in terms of BOD and that the legislation on effluent discharges and the assessment of the compliance with discharge standards are normally based on BOD. This parameter represents a good indicator of the biodegradable organic matter content of water. BOD is used to measure the quality of water in a jar: surface water (rivers, lakes, etc.) and wastewater (wastewater treatment plants, industrial discharges, etc.). It can be considered as an indirect indication of biodegradable organic carbon.

The COD test is increasingly used for design, mathematical modeling and performance evaluation purposes.

1.1.2.2. *Chemical oxygen demand (COD)*

Chemical oxygen demand (COD) is the measure of the amount of oxygen required to chemically oxidize the oxidizable organic and inorganic matter contained in a sample. COD is measured using an energetic oxidant such as potassium dichromate in an acidic medium, maintained hot for two hours. It is estimated that this oxidation destroys 90–95% of the compounds. This parameter gives an estimate of the amount of pollutants present in the wastewater. The value obtained is therefore an indirect indication of the level of organic matter present. It is one of the most widely used methods for assessing the overall load of organic substances.

COD measures all oxidizable substances, whether they are biodegradable or not (persistent organic pollutants or organometallics). The main difference between these two tests is that BOD is related to the biochemical oxidation of organic matter achieved entirely by microorganisms, whereas COD corresponds to the chemical oxidation of organic matter, obtained by a powerful oxidant (potassium dichromate) in an acidic medium.

The main benefits of the COD test are:

– The test only lasts two hours.

– Due to its quick fast response, the test can be used for operational control.

– The test results provide an indication of the oxygen required for the stabilization of organic matter.

– The test makes it possible to establish stoichiometric relationships with oxygen.

– The test is not affected by nitrification, giving an indication of the oxidation of carbonaceous organic matter only.

The main limitations of the COD test are:

– The biodegradable and inert fractions of organic matter are oxidized. Consequently, the test may overestimate the oxygen needed during biological wastewater treatment.

– Some reduced inorganic constituents could be oxidized and interfere with the result.

However, COD also applies to compounds that play no role in the oxygen deficit in a waterway. These are, for example, mineral halides (except fluorine). It is often recommended that measuring the COD be performed before the BOD, in order to estimate the proportion of dilutions to be performed.

In domestic wastewater, the COD/BOD_5 ratio varies between 1.7 and 2.4. Depending on the value of this report, conclusions can be drawn about the biodegradability of wastewater and the treatment process to be implemented (Braile and Cavalcanti 1979):

– a COD/BOD_5 ratio smaller than 3: the biodegradable fraction is high; good indication of biological treatment;

– a high COD/BOD_5 ratio (greater than 3.5 or 4): the inert fraction (non-biodegradable) is high, justifying the interest of a physicochemical process.

For non-agricultural uses, a biological treatment may be used for nitrogen removal if it could not be completely treated at the treatment plant.

Table 1.3 gives some values of the BODs and CODs of some organic products found in water, expressed in mg $O_2.g^{-1}$ of product.

Organic substance	O_2 theoretical consumption	BOD_5	BOD_{20}	COD
Acetamide	1,080	745	–	–
Acetone	2,210	544	–	2,210
Acetophenone	2,532	518	–	2,532
Adipic acid	1,423	598	1,115	1,337
Acetic acid	1,066	700	900	1,066
Propionic acid	1,513	1,300	1,400	1,497
Valeric acid	2,039	1,400	1,900	1,957
Citric acid	686	420	610	–
Amylic acid	2,727	1,265	1,732	–
Benzoic acid	1,967	1,250	1,400	1,967
Succinic acid	1,305	419	607	1,305

Organic substance	O_2 theoretical consumption	BOD_5	BOD_{20}	COD
Butyric acid	1,818	1,150	1,450	1,808
Fumaric acid	827	175	–	818
Gallic acid	1,320	80	–	–
Glutaric acid	1,212	433	1,241	–
Uric acid	667	300	–	–
Lactic acid	1,067	540	960	1,067
Oxalic acid	178	100	115	178
Phthalic acid	1,444	1,000	–	–
Salicylic acid	1,623	950	1,250	1,623
Undecanoic acid	2,879	537	–	1,986
Tartaric acid	533	350	460	522
Stearic acid	2,929	117	–	878
Palmitic acid	2,875	58	–	SO5
Myristic acid	2,807	56	–	842
Capric acid	2,604	234	–	2,213
Methyl alcohol	1,500	960	1,260	1,485
Ethyl alcohol	2,090	1,350	1,800	2,027
Benzyl alcohol	2,519	1,550	1,950	2,418
Butyl alcohol	2,594	1,505	1,885	2,403
Octyl alcohol	2,950	1,088	–	2,891
Lauric alcohol	3,096	619	–	1,269
Cetyl alcohol	3,173	–	–	1,075
Hexanol	2,823	790	–	2,653
Phenol	2,383	1,700	2,000	2,383
Naphthol	2,542	932	1,598	2,313
Benzene	3,076	377	–	1,156
Glycerol	1,217	307	–	584
Resorcinol	1,890	1,150	–	–
Lactose	1,066	580	860	–
Glucose	970	580	860	850
Starch	1,185	680	900	–
Peptone	1,407	149	–	–

Organic substance	O_2 theoretical consumption	BOD_5	BOD_{20}	COD
O-cresol	2,519	1,600	1,800	–
Hydroquinone	1,890	478	–	1,871
Glycol	1,290	400	1,080	1,150
Glycerine	1,217	720	940	–
Dextrin	1,185	520	840	–
Aniline	2,660	1,760	–	–

Table 1.3. *BOD and COD of a few organic substances in $mg.g^{-1}$ of product*

If all organic matter is biodegradable, we must have:

$$COD/BOD_{20} = 1$$

If the material is not all biodegradable, we have:

$$COD/BOD > 1$$

The theoretical oxygen demand of a Cx Hy Oz product is as follows:

$$C_x H_y O_z + nO_2 \rightarrow x\ CO_2 + y/2\ H_2O$$

The theoretical oxygen demand in g of oxygen per mole of product is:

$$32\ ((4x + y - 2z)/4)\ g\ O_2/mole$$

1.1.2.3. *Total organic carbon (TOC)*

To be more accurate with treated wastewater, the recommendation is to use total and dissolved organic carbons (TOC and DOC) which are much more reliable. Organic carbon can even be measured in-line and in situ. It indicates the total amount of organic matter expressed as carbon ($mg.L^{-1}$) contained in a solution. It makes it possible to directly know the organic load of a water. Dissolved organic carbon (DOC) represents the organic matter remaining after filtration on membranes of 0.45 micrometers.

In wastewater reuse and depending on how the water produced will be used, the total organic carbon is an important parameter used for monitoring the overall levels of the existing organic compounds. It reflects the importance of having a general indicator of the approximate level of organic contamination that is easy to measure.

The most common measurement principle consists of burning the sample in a pure oxygen atmosphere and then directing the combustion gases towards an analyzer connected to a recorder, the response of which is proportional to the amount of organic carbon present. In general, the device is calibrated with solutions of pure products, at concentrations close to the sample to be analyzed. The main combustion gas is CO_2. It is formed by the combustion of organic products, after being conveyed through a tube furnace at 800–900°C.

When analyzing TOC, the following elements are measured:

– TC = total carbon;

– TIC = total inorganic carbon (bound to carbonates);

– POC = purgeable organic carbon;

– NPOC = non-purgeable organic carbon;

– DOC = dissolved organic carbon;

– POC = particulate organic carbon.

Volatile organic compounds have a low boiling point and can be purged from a solution by injecting gas into a sample.

In summary, we have:

– TOC = TC – TIC;

– TOC = POC + NPOC.

Generally speaking, the following relation is used:

$$TOC = DOC + POC$$

After membrane filtration during which all POC is removed, it follows that:

$$TOC = DOC$$

This principle is also used after on-material filtration when turbidity is < 2 NTU. DOC is found in colloids, mainly humic acids, fulvic acids, and small molecules resulting from the decomposition of organic matter (proteins, lipids, carbohydrates). The DOC is the most widely used indicator for designing, mathematical modeling and performance evaluation.

Generally speaking, the term "particulate" is used to indicate that solids are present in the form of suspended solids. Within this context, terms such as particulate BOD, COD, phosphorus, etc., are used to indicate that they are related to suspended solids. On the other hand, soluble BOD, COD and phosphorus are associated with dissolved solids.

This test is mainly used in research or in detailed evaluations of wastewater characteristics, due to the high cost of equipment.

1.1.3. *Nitrogen*

In its cycle in the biosphere, nitrogen alternates between various forms and states of oxidation, resulting from various biochemical processes (Table 1.4).

Form	Formula	Oxidation state
Molecular nitrogen	N_2	0
Organic nitrogen	Variable	Variable
Free ammonia	NH_3	–3
Ammonium ion	NH_4^+	–3
Nitrite ion	NO_2^-	+3
Nitrate ion	NO_3^-	+5

Table 1.4. *Different forms and states of nitrogen oxidation*

Nitrogen is a component of great importance in terms of water pollution control, for the following reasons:

– Water pollution:

- Nitrogen is an essential nutrient for algae leading, under certain conditions, to the phenomenon of eutrophication of lakes and reservoirs.

- Nitrogen may result in the consumption of dissolved oxygen in the receiving medium due to processes of conversion of ammonia into nitrites and nitrites into nitrates.

- Nitrogen in nitrate form is associated with diseases such as methemoglobinemia.

– Wastewater treatment:

- Nitrogen is an essential nutrient for microorganisms responsible for wastewater treatment.

- Nitrogen, in processes of conversion of ammonia into nitrites and nitrate into nitrites (nitrification), which can occur in a WWTP, leads to oxygen and alkalinity consumption.

- Nitrogen is involved in the process of converting nitrate into nitrogen gas (denitrification).

– In raw domestic wastewater, the predominant forms are organic nitrogen and ammonia. Organic nitrogen corresponds to the amino groups. Ammonia mainly originates from urea, which is rapidly hydrolyzed and rarely found in raw wastewater. These elements are determined in the laboratory by the Kjeldahl method, leading to total Kjeldahl nitrogen (KTN).

In brief:

$$KTN = NH_3 + \text{organic nitrogen}$$

$$NT \text{ (total nitrogen)} = KTN + NO_2^- + NO_3^- = NH_3 + \text{organic nitrogen} + NO_2^- + NO_3^-$$

The distribution of ammonia in wastewater can be schematically represented in Figure 1.5. It can be seen that the fraction of oxidized nitrogen NOx (nitrite + nitrate) is very low, or even non-existent in raw wastewater. KTN can be subdivided into a soluble fraction (dominated by ammonia) and a particulate fraction (associated with suspended organic matter).

Ammonia exists in solution as an ion (NH4+) and in free non-ionized form (NH3), according to the following dynamic equilibrium:

$$NH_3 + H^+ \leftrightarrow NH_4^+$$

Free ammonia *Ionized ammonia*

At the output of the treatment plants, several forms of nitrogen can therefore be found:

– oxidized nitrogen: nitrites and nitrates;

– non-oxidized forms: Kjeldahl nitrogen including organic nitrogen and ammoniacal nitrogen (NH_4^+);

– ammonifiable organic nitrogen;

– refractory organic nitrogen.

Depending on the country's regulations, the residual form of nitrogen may be ammonical if the discharge standards do not enforce a limit value for this parameter. As a result, its presence in organic or ammoniacal form results in oxygen consumption in the natural environment and modifies the living conditions of the environment.

In France, where nitrogen regulations are rather strict, nitrification and denitrification must be optimally controlled.

Total nitrogen includes Kjeldahl nitrogen (KTN KN), nitrites and nitrates (see Figure 1.5).

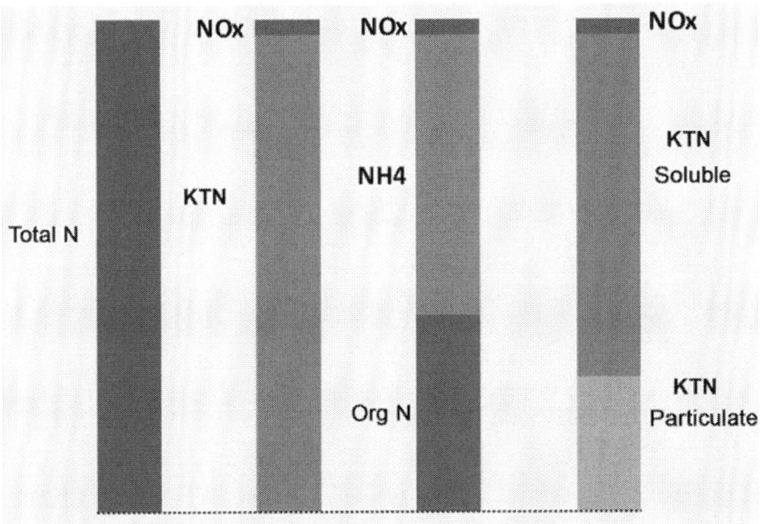

Figure 1.5. *Distribution of nitrogen forms in untreated domestic wastewater (IAWQ 1995)*

The relative distribution introduces the following values, depending on pH values:

– pH < 8: nearly all ammonia is in NH_4^+ form;

– pH = 9.5: about 50% NH_3 and 50% NH_4^+;

– pH > 11: nearly all ammonia is in NH_3 form.

It can be seen that, in the usual pH range, close to neutrality, the ammoniac existing in wastewater is practically in ionized form. Water temperature also influences this distribution. At a temperature of 25°C, the proportion of free ammonia to total ammonia is about twice as high compared with a temperature of 15°C.

The following formula can be used to calculate the proportion of free ammonia relative to total ammonia present in water, as a function of temperature and pH (Emerson et al. 1975):

$$\frac{\text{free NH3}}{\text{total NH3}}\,\% = \left(1 + 10^{0.09018+\left(\frac{2,729.92}{T+273}\right)-\text{pH}}\right)^{-1} \text{x } 100$$

where T = wastewater temperature (°C).

Table 1.5 and Figure 1.6 show the variation in the proportions of free NH_3 and NH_4^+.

pH	T° = 15°C		T° = 20°C		T° = 25°C	
	% NH_3	% NH_4^+	% NH_3	% NH_4^+	% NH_3	% NH_4^+
6.50	0.09	99.91	0.13	99.87	0.18	99.82
7.00	0.27	99.73	0.40	99.60	0.57	99.43
7.50	0.86	99.14	1.24	98.76	1.77	98.23
8.00	2.67	97.33	3.82	96.18	5.38	94.62
8.50	7.97	92.03	11.16	88.84	15.25	84.72
9.00	21.50	78.50	28.43	71.57	36.27	63.73
9.50	46.41	53.59	55.68	44.32	64.28	35.72

Table 1.5. *Proportion of free and ionized NH_3 as a function of pH and temperature*

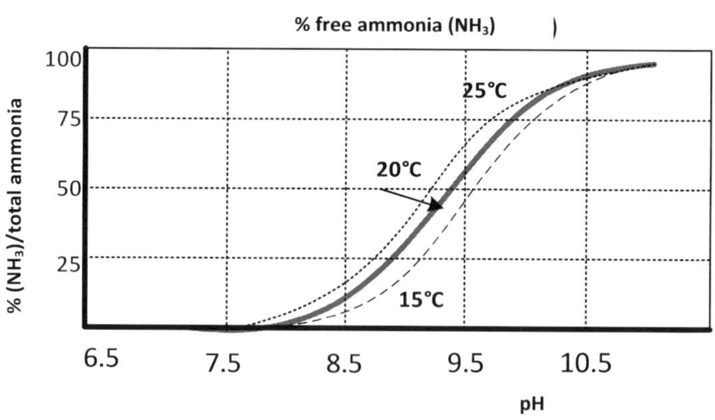

Figure 1.6. *Percentage of free NH_3 as a function of pH and temperature*

1.1.4. *Phosphorus*

Total phosphorus in domestic wastewater is present in phosphate form, classified as follows:

– inorganic (polyphosphates and orthophosphates): main source of detergents and other household chemicals;

– organic (bound to organic compounds): physiological origin.

Phosphorus in detergents is present, in raw wastewater, in the form of polyphosphates or, after hydrolysis, in orthophosphate form.

Orthophosphates are directly available to the biological metabolism. The forms in which orthophosphates are present in water depend on the pH and include PO_4^{3-}, HPO_4^{2-}, $H_2PO_4^-$, H_3PO_4.

In wastewater, the predominant form is HPO_4^{2-}.

Polyphosphates are more complex molecules, with two phosphorus atoms or more. Polyphosphates are transformed into orthophosphates by hydrolysis, which despite its slow process, takes place in transport pipelines. Phosphorus in detergents can account for up to 50% of the total phosphorus in wastewater.

Phosphorus in wastewater can be in solid form:

– soluble phosphorus (mainly inorganic): mainly polyphosphates and orthophosphates (inorganic phosphorus), as well as a small fraction corresponding to phosphorus bound to soluble organic matter in wastewater;

– particulate phosphorus (fully organic): bound to particles of organic matter in wastewater.

The importance of phosphorus is associated with the following aspects:

– phosphorus is an essential nutrient for the growth of microorganisms. Domestic wastewater generally contains adequate levels of phosphorus;

– phosphorus is an essential nutrient for algae growth, eventually leading, under certain conditions, to the eutrophication of lakes and reservoirs. It is estimated that from 20 $\mu g.L^{-1}$, the effect of P on algae growth is noticeable.

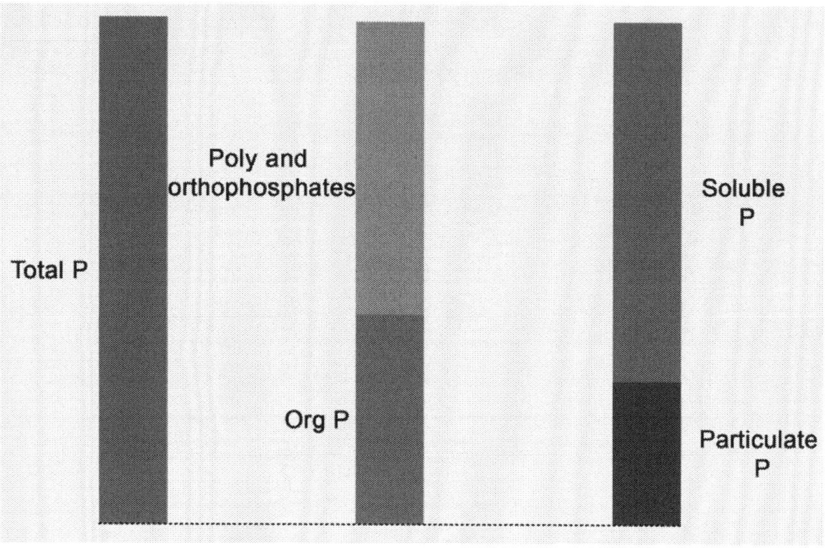

Figure 1.7. *Distribution of phosphorus forms in urban wastewater*

1.1.5. *Micropollutants*

1.1.5.1. *Pesticides and endocrine disruptors*

Over the past decade, a large number of scientific articles have quantified and described the performance of wastewater treatment plants in relation to these substances. It thus appears that treated water still contains a concentration that most often cannot be overlooked. This is expected because the equipment available to the treatment plants is not effective enough to remove these substances.

The most represented families of pesticides include:

– organophosphates (chlorpyrifos, malathion) which have an insecticidal effect by way of blocking acetylcholinesterase (disruption of the transmission of nerve impulses);

– organochlorines (DDT, chlordecone, lindane), insecticides affecting the nervous system by disrupting nerve impulses at the synapses. These compounds are now all banned in France but are widely encountered in the environment (as, for instance, in the West Indies, where chlordecone is causing problems);

carbamates (aldicarb, carbofuran), which act on thc ncrvous system in the same way as organophosphates;

– triazines (atrazine, simazine), which have herbicidal properties by blocking the transmission of electrons and light energy;

– phenylureas (diuron, isoproturon), which have a mode of operation comparable to that of triazines;

– chloroacetanilides (metolachlor, acetochlor), which act on plants by inhibiting elongases, enzymes involved in lipid synthesis (long-chain fatty acids);

– aryloxyacids (2.4 D, mecoprop, dichlorprop), which disrupt auxin AIA (indole 3-acetic acid) regulation, a phytohormone essential for plant growth;

– sulfonylureas (chlorsulfuron, nicosulfuron), which affect the synthesis of branched-chain amino acids (valine, leucine, isoleucine) through the inhibition of acetolactate synthetase.

Main families	**Compounds**
Drug molecules	Anxiolytics, analgesics, anti-inflammatories, antibiotics, lipid-lowering agents, antihypertensives, anti-epileptics, anticancer, iodinated contrast agents, etc.
Cosmetics and personal care products	Musks (galaxolide, tonalid), triclosan, etc.
Endocrine disruptors	Hormones, alkylphenols, BPA, phthalates, flame retardants, etc.
Illegal drugs	Cocaine and metabolites, THC, amphetamines, etc.
Pesticides	Insecticides, herbicides
Priority substances (Appendix X of the WFD) and hazardous substances (RSDE notification 050109)	PAH, dichloromethane
Disinfection by-products	Trihalomethanes, haloacetic acids

Table 1.6. *Main families and organic compounds*

Agricultural activity is not the only source of pesticide emissions. Among other things, secondary sources include the maintenance of public spaces, golf courses, private use, the annual control of mosquitoes in wetlands, the veterinary sector, road maintenance, etc.

1.1.5.2. *Hormones*

Natural (estradiol and its derivatives) and synthetic (ethynylestradiol in contraceptive pills) animal hormones are excreted by humans and end up in wastewater. Natural estrogen hormones comprise estrone (E1), 17α-estradiol (E2) and estriol (E3). The most important synthetic hormone, as it is used in birth control pills, is 17α-ethynylestradiol (EE2). Women of reproductive age excrete at least 50 μg.J^{-1} of estrogen hormones in their urine. This level can exceed 5 mg.J^{-1} for a pregnant woman. Men also excrete it, but in much smaller proportions, of approximately a few micrograms. Testosterone is also excreted but efficiently removed by mineralization in wastewater treatment plants (Barbier 2011).

	Plant influents		**Plant effluents**	
	Min	**Max**	**Min**	**Max**
		A		
Estradiol	0.61	51.7	0.04	19.1
Estrone	0.90	17.6	0.04	3.62
Estriol	1.11	14.9	0.22	3.14
Ethinyl E2	1.03	48.49	0.3	14.89
		B		
Estradiol	0.4	14.41	0.17	2.2
Estrone	0.13	19.8	0.06	7.75
Estriol	0.45	18	0.09	3.58
Ethinyl E2	0.77	4.09	0.3	1.47
		C		
Estradiol	0.25	36.2	0.003	2.69
Estrone	2.19	13.2	0.2	6.09
Estriol	0.8	9.75	0.17	3.65
Ethinyl E2	1.07	8.89	0.46	2.93

Table 1.7. *Estrogen concentrations in influents and effluents of a few wastewater treatment plants*

The concentrations are therefore highly variable. However, the removal rates are high, of about 70–85%, except for ethinylestradiol, where it varies from half to two-thirds.

Other products that are likely to have such effects include alkylphenol ethoxylates (APEOs), which are found in industrial cleaning products, paints and some coatings. Of these products, octylphenol and nonylphenol have an endocrine disrupting effect that has been analyzed up to concentrations as high as 10 μg.L^{-1}. Phthalates such as DEHP (dietylhexylphthalate) and BBP (butylbenzylphthalate) range from 0.2 to 1 μg.L^{-1}.

Substance	Family	Efficiency (%)	Wastewater – influents (µg.L^{-1})	Treated water – effluents (µg.L^{-1})
4-nonylphenol	Alkylphenol	84 (60–100)	15.7	1.3
Pentabromodiphenyl ether	PBDE	98 (90–100)	0.39	Nc
Atrazine	Pesticide	2 (1–30)	0.02	0.02
4-t-octylphenol	Alkylphenol	88 (70–100)	5.6	0.21
Bisphenol A	Phenolated polymer	60 (30–90)	0.16	0.06
4-NP1EO	Alkylphenol	88 (70–100)	9.0	0.47
DEHP	Phthalate	92 (80–100)	52.8	4

Table 1.8. *Efficiency of wastewater treatment plants for some endocrine disruptors (Lopez and Laurent 2013)*

Drug residues are among the main xenobiotics present plant effluents and in the environment. Not all of them are endocrinic disruptors but some are. Among the most common products, we can find carbamazepine, an anticonvulsant, and lopromides, which are contrast agents for radiology. Finally, antibiotics should also be of particular concern because they are found in all aquatic compartments and accumulate in sediments.

1.1.5.3. *Polybrominated flame-retardant compounds*

The most common are polybrominated diphenyl ethers (PBDEs), the different types of hexabromocyclododecane (HBCD) and tetrabromobisphenol A (TBBPA). These are lipophilic, bioaccumulative persistent organic substances.

1.1.5.4. *Perfluorinated compounds*

The two compounds that receive the most attention are perfluorooctane sulfonate (PFOS) and perfluorooctanoic acid (PFOA). They are hydrophobic and lipophobic.

1.1.5.5. *Parabens*

Because of their ubiquity, parabens are found in almost all wastewater.

1.1.6. *Parametric influence in the treatment system*

The characteristics of the effluents to be treated must necessarily be validated by 24-hour measurement campaigns (in dry and rainy weather). These campaigns make it possible to verify the ratios of the different parameters to each other: DCO/DBO_5; MES/DBO_5; DBO_5/KTN; N-NH_4/KTN; DCO/Pt; MVS/MES and DCO/KTN.

Unit operations will be based on these ratios (Table 1.9), and the influence of each of them will be taken into account during the treatment process.

Ratio	Urban effluent	Observation
COD/BOD_5	2.2–2.4	Indicates the influent biodegradability
TSS/BOD_5	0.8–1.2	Will affect the % MVS of the effluent and sludge production
BOD_5/KTN	4–5	Will affect the treatment of nitrogen and the conditions of the biological mechanisms
N-NH_4/KTN	0.6–0.8	Indicates the degree of ammonification of the influent
COD/Pt	45–50	Indicates the potential biological treatment of phosphorus
VSS/SS	0.65–0.75	Will influence sludge production, the design of the aerated lagoon, clarifier design and sludge treatment
COD/NTK	8–12	Influences the denitrification treatment and the interest of an anoxia zone separated from the aeration tank

Table 1.9. *Parametric influence on the treatment system*

Wastewater flow rates are 150–250 L/inhabitant, depending on the agglomeration, the quality and specificities of the distribution networks and how water is used.

The average concentrations in wastewater are 453 mg.L^{-1} of total solids, 217 mg.L of total volatile solids, 145 mg.L^{-1} of suspended solids, 120 mg.L of suspended volatile solids, 288 mg.L^{-1} of chemical O_2 demand and 147 mg.L^{-1} of biochemical O_2 demand for a pH varying from 6.8 to 7.5.

Volatile and non-volatile fatty acids account for 8.5–20 mg.L^{-1} and 0.1–1 mg.L^{-1}, respectively; higher fatty acids represent 2/3 of the fatty acid content and proteins and amino acids 45–50% of total nitrogen.

The BOD_5/N/P ratio is in average 100/5/5 with about 10% fat, 40% nitrogenous solids and 50% carbohydrates.

The minimal composition considered in price calculations is, for a PE, 60–70 g of dissolved BOD5, 120 g of COD and 70–90 g of SS (1/3 mineral solids (MS), 2/3 organic solids (MVS)). The fraction of suspended solids in raw municipal wastewater is generally about 72 g per population equivalent.

Parameters	Quantity (g)	Concentration ($mg.L^{-1}$)
Suspended solids (SS)	90	100–400
Oxidizable solids (OS)*	57	–
DBO_5	60	150–500
COD	120	300–1,000
$N\text{-}NH_4$	15	30–100
Total phosphorus	4	–
Inhibitory materials	0.2	–
METOX**	0.23	–
AOX***	0.05	–
Fat	22	–

*MO = (COD+ 2BOD)/3; **METOX: total toxic metals; ***AOX: adsorbable organohalogenated compounds on activated carbon.

Table 1.10. *Overall characteristics of domestic wastewater*

Equivalence	METOX number
1 g Zn	1
1 g Cu or Ni	5
1 g Pb or As	10
1 g Hg	50
1 g Cd	50

Table 1.11. *METOX equivalence of some heavy metals*

The legislation defines the concept of population equivalent as "the biodegradable organic load with a five-day biochemical oxygen demand (BOD_5) of 60 grams of oxygen per day" (European Directive 91-271).

1.2. Microbiological parameters: pathogenic organisms and indicators of fecal contamination

The role of bacteria as testimony of pollution has been known for a long time as various links have been established between water contamination, the presence of bacteria and diseases registered in individuals in the immediate vicinity of the discharge area. It has been estimated that every human being eliminates several billion germs per gram of faces every day, including 100–500 million coliforms. Therefore, in patients with enteritic disease, pathogenic bacteria reach high levels in the stool (250,000 to several billion salmonellae per gram). In paratyphics, the figures recorded are much higher and can reach several hundred billion per gram of fecal matter.

Most microorganisms play a variety of roles, including the transformation of constituents in biogeochemical cycles. Another important aspect in terms of biological wastewater quality is that related to the transmission of diseases by pathogenic organisms.

The groups of pathogenic organisms are:

– bacteria;

– viruses;

– protozoa;

– helminths.

A water-related disease is defined as any adverse, significant or widesperad effect on humans, such as death, disability, disease or disorder, directly or indirectly caused.

It is in the digestive system that the first reactions are born, in particular the development of digestive parasitosis, of which there are two main families: protozoa and metazoans. Protozoa include amoebas, flagellates and ciliates. Metazoans include cestodes (or tænia), nematodes and trematodes.

Another category of bacteria comprise mycobacteria, responsible for tuberculosis. Koch's bacillus is eliminated through the feces and is found in particular in wastewater discharged by sanatoriums and specialized hospitals. The population of Koch's bacilli can be very large in some cases: 10,000 germs per ml of water, or 100,000 bacilli per ml of sludge.

There are also many other bacteria such as those responsible for abscesses (flavobacterium, chromobacterium, semitia) and also for cholera germs or clostridium-type bacteria (brought by rainwater runoff on contaminated crop fields).

Wastewater also contains viruses present in stools of healthy, sick or recovering carriers. The main viruses found belong to the enterovirus group, which are located in the human intestine (poliovirus, echo, reovirus), or to virus A of viral hepatitis. Polioviruses can reach a concentration of 100,000 units per gram in feces, and the figures are of the same order of magnitude for other viruses, although not all authors agree on the significance of these values.

Wastewater also carries plant and animal parasites and parasite eggs and cysts. Various analyses conducted have highlighted the presence, for example, of saccomyces, torula, candida, trichosporon, ascomycetes, etc.

The quantities of these parasites greatly vary and fall within a very broad range (10,000–100,000 colonies per liter of water). Microbiological pollution becomes very dangerous when wastewater is discharged into a receiving environment that can be used as a leisure spot (seawater, lake, river, etc.).

There are a large number of diseases that can be contracted, and the most typical are ocular diseases, nasopharyngeal and optic disorders, cutaneous manifestations, impetigos, furunculosis and mycosis.

Intestinal diseases and diseases due to viruses should also be included. Consequently, we can be certain that microbiological pollution is serious, insidious and a constant nuisance for the individual, in particular when wastewater is discharged upstream or into environments likely to be recreational places.

Table 1.12 provides a limited inventory of the origins of and diseases caused by organisms in wastewater.

Organism	Description	
Bacteria	Unicellular organisms. Existing in different shapes and sizes. Responsible for the degradation of organic solids and nitrogen products in wastewater treatment plants. Responsible for waterborne or food-borne diseases.	
	E. coli (enterotoxin)	Diarrhea
	Leptospira	Fever, epidermal hepatitis, meningitis
	Salmonella	Gastroenteritis, vomiting
	Shigella	Colitis, colic, diarrhea
	Yersinia enterocolitica	Fever, enteritis, erythema
	Pseudomonas aeruginosa (pyocyanic bacillus)	Urinary tract inflammation
	Staphylococcus	Abscess, airways inflammation
	Streptococcus	Local infection, endocarditis, otitis
	Klebsiella pneumoniae	Pneumonia

Virus	Pathogenic and difficult to eliminate during water treatment. Includes enteroviruses, viruses hepatitis A, Norwalk-type gastroenteric virus, reovirus and rotavirus, adenovirus.
Protozoans Enteric	Some enteric protozoa are pathogenic and have been associated with outbreaks of drinking water-related diseases. The protozoa of greatest concern are *Giardia* and *Cryptosporidium*. These organisms can be found in water as a result of direct or indirect contamination with human or animal feces. The transmission of *Giardia* and *Cryptosporidium* often occurs on a person-to-person basis.
Helminths	Parasitic worms in the digestive tract of humans or animals; they release a large number of eggs with a shell consisting of several membranes of varying thickness into the external environment. This resistant structure allows them to achieve the biological cycle and to perpetuate their infesting power. These microscopic eggs vary in size from 30 to 160 μm. Intestinal helminthiases are animal and human parasitic diseases caused by helminth infestation. These pathogens can be transmitted to humans through direct contact with wastewater, indirectly through the consumption of crops irrigated with this wastewater, or through food products of animal origin.

Table 1.12. *Most frequently analyzed microorganisms in wastewater*

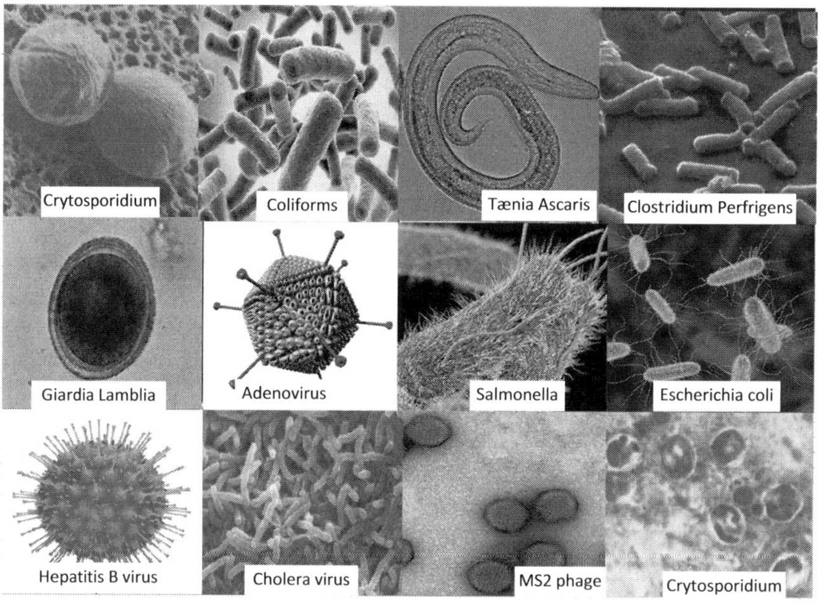

Figure 1.8. *Some microorganisms found in wastewater*

1.2.1. *Bacteria*

1.2.1.1. *Coliforms*

The most commonly mentioned microorganisms are bacteria of the coliform group. These are present in large quantities in human stool. Each individual excretes an average of 10^{10}–10^{11} cells per day (Branco 1979).

About 1/3–1/5 of the weight of human stool is made up of coliform bacteria. In wastewater, the concentration of coliforms is much higher than that of other pathogenic bacteria. In wastewater, coliforms have a higher resistance than the majority of pathogenic enteric bacteria. This is an important characteristic and one of the reasons why they are a good indicator of fecal contamination.

The coliform group of bacteria consists mainly of species of the genera *Citrobacter*, *Enterobacter*, *Escherichia* and *Klebsiella* and includes fecal coliforms where *Escherichia coli* is the predominant species. Many of the coliforms are capable of growing outside the gut, especially in warm climates; as such, their number is not a suitable measure to be used as a monitoring parameter for wastewater reuse systems. The fecal coliform test can also include some non-fecal organisms that can grow at 44°C; therefore, the number of *E. coli* is the most satisfactory indicator parameter for wastewater use in agriculture.

Coliform removal mechanisms in treatment plants are the same as those used for pathogen bacteria. As such, the removal of pathogenic bacteria is usually associated with the removal of coliforms. However, other pathogenic organisms (such as protozoan cysts and helminth eggs) can be removed by different mechanisms.

Bacteriological techniques for the detection of coliforms are fast and affordable compared to those of pathogens.

1.2.1.2. *Fecal streptococci*

The group of fecal streptococci includes two main genera: *Enterococcus* and *Streptococcus*. The genus *Enterococcus* comprises many species, most of them of fecal human origin; however, some species are of animal origin. All *Enterococcus* have a high tolerance to adverse environmental conditions. Fecal streptococci rarely multiply in polluted waters and are more resistant than *E. coli* and coliform bacteria. For these reasons, they are sometimes used as indicators for bathing waters (WHO 2006).

The relation between fecal coliform values and fecal streptococci (FC/FS ratio) is used in some countries to give an indication of the origin of contamination, whether mainly human or animal. High FC/FS values would suggest mainly human contamination, while low FC/FS values would suggest mainly animal contamination. These are special cases despite the fact that these relationships are not applicable in every case, and may give wrong indications of contamination in various reservoirs. This group of organisms comprises species primarily associated with animals (*Streptococcus bovis* and *S. equinus*), other species with a wider distribution (such as *S. faecalis* and *S. faecium*, present in humans and other animals) as well as two biotypes (*S. faecalis var liquefaciens* and a typical *S. faecalis* that hydrolyzes starch) that appear to be ubiquitous, occurring in both polluted and unpolluted environments. Counting the number of fecal streptococci in effluents is a simple routine procedure but it has the following limitations: 1) the possible presence of non-fecal biotypes in the natural microflora of crops may affect their usefulness in assessing the bacterial quality of crops irrigated with wastewater; 2) the lower survival of fecal streptococci at high temperatures than at low temperatures. Further studies are still needed on using fecal streptococci as an indicator in tropical conditions and especially for comparing survival with that of salmonella.

1.2.1.3. *Sulfito-reducing clostridia*

Clostridium perfringens are the most representative species of this group, normally occurring in feces. The concentration of *Clostridium perfringens* is lower than that of *E. coli*. This bacterium is a sporulated anaerobe, exclusively fecal, normally used for detecting intermittent or past water pollution, due to the prolonged survival of its spores. Although this lengthy survival is generally considered unsuitable for normal purposes, it can be very useful in wastewater reuse studies, because *Clostridium perfringens* may prove to have survival characteristics similar to viruses or even helminth eggs.

Clostridium spores can survive in water much longer than organisms in the coliform group and are more resilient to disinfection. Their presence in disinfected water may indicate treatment failure.

Pathogen bacterial concentration in treated wastewater falls between 10^6 and 10^7 bacteria per 100 mL^{-1}. That of enterobacteriaceae is about 10^5–10^6 u.100 mL^{-1}, with 10^3–10^4 for streptococci and 10^3–10^4 100 mL^{-1} for clostridia.

1.2.1.4. Salmonella *spp.*

Several species of *Salmonellae* may be found in the raw wastewater of an urban community in a tropical developing country, including *S. typhi* (typhoid fever causation agent) and many others. It is estimated (Doran et al. 1977) that a count of 7,000 *Salmonellae*/liter is typical in a tropical urban sewer with a similar number of Shigella, and possibly 1,000 *Vibrio cholera*.L^{1}. *Shigella* spp. and *V. cholera* are both more quickly killed in the environment, therefore if salmonella can be removed, then the majority of other bacterial pathogens will have been removed as well.

1.2.2. *Enteric viruses*

1.2.2.1. *Enterovirus*

About 140 types of enteric virus diseases have been described and are potentially present in wastewater. All viruses are pathogenic to humans and cause a wide spectrum of diseases and symptoms. Let us mention enteroviruses such as polioviruses, reoviruses, rotaviruses, adenoviruses, astroviruses, caliciviruses such as Norwalk's agents, and hepatitis A and E viruses. Enteric viruses are present in smaller numbers than bacteria in wastewater and do not multiply in this environment. Nonetheless, they usually have a low infective dose and can survive for several months in water under certain conditions.

The concentration of enteroviruses in treated wastewater is between 10^3 and 10^4 u.L^{-1}.

Enteric virus diseases are human pathogens most likely to contaminate groundwater (irrigation, groundwater recharge), especially since their extremely small size (20–100 nm) enables them to infiltrate soils, eventually reaching aquifers. Depending on the physicochemistry and virus-specific factors (such as size and isoelectric point, for example), viruses can easily travel in soils and persist for several months in soils and subsoils.

Enteroviruses have also proved to be more resistant to disinfection than bacterial indicators. They give rise to serious diseases, such as poliomyelitis and meningitis, or to a range of minor diseases such as respiratory infections. Although there is no strong epidemiological evidence that these diseases spread through wastewater irrigation systems, there is some risk and we should know to what extent viruses are removed by existing and new treatment processes, especially under tropical conditions. Virus counts can only be performed in a dedicated laboratory, since the required cell culture techniques are highly sensitive to bacterial and fungal contamination.

Pathogen protozoa and enteric viruses are considered priority microbial contaminants because of their important role in waterborne disease outbreaks and associated public health risks, their lengthy survival in the environment, their resilience to conventional water disinfection processes and the often weak or non-existing correlation with the traditional bacterial water quality indicator.

1.2.2.2. *Rotaviruses*

These viruses are known to cause gastrointestinal problems and, although generally present in fewer numbers than enteroviruses in wastewater, they are known to be more persistent; it is therefore necessary to establish their survival characteristics in relation to the enterovirus and relatively to indicating organisms in wastewater. It has been claimed that the removal of viruses in wastewater treatment occurs in parallel with the removal of suspended solids, because most viral particles are associated with solids. Consequently, suspended solids must be measured in treated effluents as a routine.

1.2.2.3. *Bacteriophages*

Bacteriophages can be representative of the presence of viruses, due to their similarities to human enteric viruses. Bacteriophages are specific viruses that infect bacteria, such as, for example, coliphages, which infect *E. coli*. The number of coliphages can be greater in wastewater, due to their fast reproduction rate resulting from the aggression of bacterial cells. They can be representative as indicators of wastewater contamination and, because of their greater persistence compared to bacterial indicators, and also as additional indicators of the effectiveness of the treatment or protection of groundwater.

1.2.3. *Enteric protozoa*

Enteric protozoa are found in wastewater in forms that are highly resistant to environmental conditions and disinfection methods. Similar to viruses, they do not reproduce outside of their host and present a low infective dose compared to bacteria. *Cryptosporidium* and *Giardia* protozoa are among the main agents responsible for diarrheal diseases in humans. The protozoa most commonly infecting humans are *Giardia lamblia*, *Entamoeba histolytica*, *Cryptosporidium*, *Cyclospora*, *Microsporidia* and others. Their resistant forms (8–13-µm *Giardia* cysts and 4–6-µm *Cryptosporidium* oocysts) are excreted in large numbers by infected animals or humans and are ubiquitous in surface waters. Enteric protozoa are resistant to harsh environmental conditions and chemical disinfectants commonly used in water treatment plants to reduce bacterial contamination. Several studies have found little or no correlation between bacterial fecal inflammation and protozoan concentration. These observations highlight the need for regular monitoring of reused water and the implementation of technologies adapted to the removal of these microorganisms for effective user protection.

1.2.4. *Helminth eggs*

Helminths do not replace bacterial indicators. However, nematode eggs such as *Ascaris*, *Trichuris* or *Necator americanus* can be used as indicators of other helminths (cestodes, trematodes and other nematodes), which are removed in water and wastewater treatment by the same mechanism (e.g. sedimentation), thus being indicators of treatment effectiveness. Helminth nematodes such as *Roundworms lumbricoides* and *Trichuris trichiura*, flatworms, cestodes such as *Tænia saginata* and trematodes such as *Schistosoma mansoni* are also microorganisms potentially pathogenic to humans that can be found in wastewater. The result of being exposed to the pathogenic microorganisms previously mentioned depends on several factors related to the host (susceptibility and degree of immunity), the pathogenic microorganism and the environment.

Helminth eggs are an important parameter when evaluating the use of treated wastewater for irrigation, in which workers may have direct contact with contaminated water and consumers may consume raw or unpeeled vegetables.

There are no surrogate indicators for helminth eggs, and these are directly determined during laboratory tests.

Helminth eggs can be eliminated with physical operations, such as sedimentation, which take place, for example, in stabilization ponds. Eggs may or may not be viable, and viability may be altered by specific disinfection processes.

Table 1.13 shows the averages observed with the bacteriological concentrations in the effluents of treatment plants (conventional activated sludge system).

Microorganism	**Concentration (u.100 mL^{-1})**
Total coliforms	10^6–10^7
Fecal coliforms	10^5–10^6
Streptococci	10^3–10^4
Enterococci	10^3–10^4
Clostridiums perfringens	10^3–10^4
Enteroviruses	10^2–10^3
Cryptosporidium parvum oocysts	10^2–10^4
Giardia lamblia cysts	10^2–10^4
Helminth eggs	0–10^2

Table 1.13. *Average bacteriological concentrations in wastewater treatment plant effluents (conventional activated sludge system)*

These values are average concentrations observed at different wastewater treatment plants around the world. However, they can be different depending on seasons when the pollutant flow is significant such as when animals are slaughtered, or a reduction in the flow rate for a higher pollutant load, etc. It should be noted that to achieve a decrease of these parameters, physical (membranes) or chemical barriers will be necessary.

1.2.5. *Fecal contamination indicators: assessment in treated wastewater*

Given the great diversity of pathogen microorganisms present in water and the limitations of detection methods, acceptable levels in treated wastewater are instead established for microorganisms indicating fecal or treatment contamination.

The important criteria that should be considered when selecting an indicator microorganism (Government of Canada 2021) are:

– The fecal pollution indicator organism must be present only if there is fecal contamination and it must not reproduce outside its host.

– The survival of the indicator microorganism to treatment and in the environment must be equal to or greater than that of the target microorganism.

– Detection and counting should be fast and inexpensive.

– The indicator microorganism must be non-pathogenic to humans.

No indicator available at the moment meets all these criteria, which is why different microorganisms and parameters are used as indicators for specific purposes.

Indicator organisms are used as indicators of the efficiency of the removal of pathogen agents in the wastewater treatment process. The pathogenic organisms that can be represented are bacteria and viruses, since they are eliminated by the same mechanisms as coliform bacteria. Protozoan cysts and helminth eggs, which are mainly removed by physical mechanisms, such as sedimentation and filtration, are not adequately represented by coliform bacteria as indicators of the efficiency of the treatment. There are various other indicator organisms, each with its advantages and disadvantages.

The most commonly used indicators of fecal contamination are: total coliforms (TC), fecal coliforms (FC) or thermotolerant coliforms, namely *Escherichia coli* (EC).

Fecal coliforms (FC) are a group of bacteria originating primarily from the intestinal tract of humans and other animals. This group includes the genus *Escherichia* and, to a lesser extent, *Klebsiella*, *Enterobacter* and *Citrobacter* species.

For decades, bacteria (e.g. fecal coliforms and *E. coli*) were used to identify fecal contamination and indicate the likely presence of microbial pathogens in the water. However, their concentrations are seldom well correlated with those of pathogens.

Therefore, bacterial indicators can signal the likely presence of pathogens in the water, but they cannot accurately predict their level of occurrence. Compared to viruses and protozoa, bacterial indicators are less persistent in the aquatic environment and less resilient to disinfection mechanisms and other treatment processes.

The total coliform group (TC) is a large group of bacteria that have been isolated in water samples and in polluted and unpolluted soils and plants, as well as in the feces of humans and other warm-blooded animals. This group has been widely used in the past as an indicator and continues as such in some areas, although issues related to the emergence of non-fecal bacteria are a problem (Thomann and Mueller 1987). Fecal coliforms (FC) are a group of bacteria primarily originating from the intestinal tract of humans and other animals. This group includes the following genera: *Escherichia* and, to a lesser extent, *Klebsiella*, *Enterobacter* and *Citrobacter* species. The FC test is performed at high temperatures, allowing the removal of bacteria of non-fecal origin. However, even under these conditions, the presence of non-fecal (freely living) bacteria is possible, although in lower numbers than total coliforms.

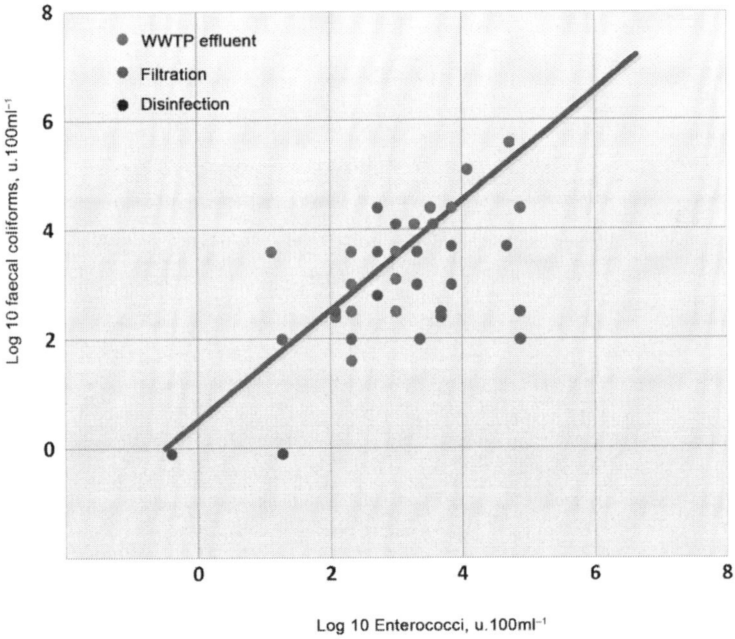

Figure 1.9. *Fecal coliforms/Enterococci correlation. For a color version of this figure, see www.iste.co.uk/gaid/wastewaterengineering1.zip*

In wastewater, *E. coli* is the predominant organism within the fecal coliform group (thermotolerant) and fecal coliforms (thermotolerant) are the predominating coliforms within total coliforms.

Therefore, even a positive fecal coliform test does not guarantee that the contamination is actually fecal. For this reason, fecal coliforms have recently been referred to as thermotolerant coliforms because they are resistant to the high testing temperatures, but are not necessarily fecal. *E. coli* is the only bacterium that guarantees contamination that is exclusively fecal. As such, there is a current trend to mainly use *E. coli* as an indicator of fecal contamination. Nonetheless, its detection does not guarantee that the contamination is of human origin, since *E. coli* can also be found in the feces of other animals. Its detection in the laboratory is very simple, established mainly by fluorogenic methods.

There is no quantifiable link between TC and pathogen microorganisms (Rose et al. 2004). However, their presence can mean the presence of other pathogen bacteria.

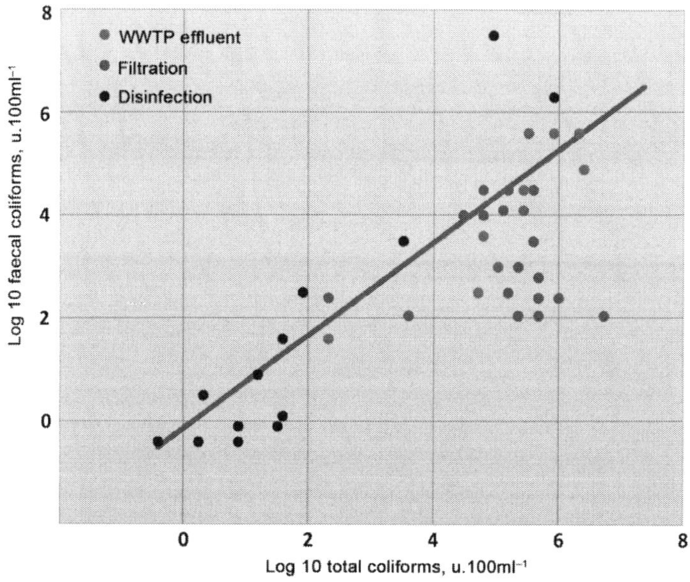

Figure 1.10. *Fecal coliforms/total coliform correlation. For a color version of this figure, see www.iste.co.uk/gaid/wastewaterengineering1.zip*

Coliforms are not good indicators of the presence of viruses or parasites (protozoa and helminths), which would be more persistent in the environment (50% reduction time in surface water concentration: coliforms = 0.9 days, cryptosporidium = up to 150 days) and more resistant to some treatments such as disinfection and have lower infective doses. To this end, other indicators such as coliphages (for viruses), *Clostridium perfringens* (for viruses and parasites), enterococci (for viruses and parasites) and a physicochemical parameter such as turbidity (for parasites) are suggested. Some countries or international organizations have added, in addition to the criteria for coliforms, criteria to be respected for parasites and viruses in reused wastewater. Despite their limitations, it should be observed that fecal coliforms and *Escherichia coli* have several advantages and are the microorganisms most widely utilized as indicators at the present time.

Microorganisms	Pollution indicators
Escherichia coli	Fecal pollution indicator Potential presence of enteric pathogens
Enterococci	Fecal pollution indicators Indirect probability of the presence of enteric viruses
F-RNA phages	Indicators of enteric virus contamination Indicators as substitutes for viruses to test the effectiveness of a process
B. fragilis phages	Fecal contamination indicators Used as microbial source monitoring parameters
Clostridium perfringens sp.	Fecal pollution indicator Indicator as a substitute for viruses and protozoa (cysts and oocysts) to test the effectiveness of a disinfection process

Table 1.14. *Examples of contamination indicators and effectiveness monitoring*

As process indicators, phage groups are often successfully used as substitutes for enteroviruses for the evaluation of the efficiency of water treatment processes. This is the case for filtration and disinfection (Rose et al. 2004).

As tracers, bacteriophages are often used both in the field and in the laboratory as biocolloids for estimating the fate and transport of surface and groundwater pathogen viruses. This use of phages as substitutes for the transport of pathogen agents applies to the evaluation of the potential health risk of pathogens in groundwater and to the evaluation of the effectiveness of treatment processes for the removal of pathogens in water.

Table 1.15 shows the values obtained in influents and effluents of wastewater treatment plants in France, where discharges are made into receiving environments located in sensitive areas.

Quality parameter (mg.L^{-1})	WWTP influent	WWTP effluent	
		Sensitive area 1	Sensitive area 2
Biochemical O_2 demand, BOD_5	200–300	15	9
Chemical O_2 demand, COD	400–600	55	39
Suspended solids	250–400	19	12
$N\text{-}NH_4^+$	35–50	3	2
$N\text{-}NO_3^-$	–	8	4
$P\text{-}PO_4^{3-}$	8–20	2	1.5

Table 1.15. *Data from a few factories on treatment performance in France*

Microorganisms	Load per capita (Org.hab^{-1}.j^{-1})	Concentration (Org/100 mL)
Total coliforms	10^{10}–10^{13}	10^{7}–10^{10}
Fecal coliforms (thermotolerant)	10^{9}–10^{12}	10^{6}–10^{9}
E. coli	10^{9}–10^{12}	10^{6}–10^{9}
Fecal streptococci	10^{7}–10^{10}	10^{4}–10^{7}
Protozoan cysts	$< 10^{7}$	$< 10^{4}$
Helminth eggs	10^{3}–10^{6}	10^{0}–10^{3}
Virus	10^{5}–10^{7}	10^{2}–10^{4}

Table 1.16. *Estimation of microorganism concentration in urban wastewater*

1.3. References

Barbier, G. (2011). Perturbateurs endocriniens, le temps de la précaution. Senate Parliamentary Office Report, no. 765, filed 12 July, Paris.

Birks, R., Colbourne, J., Hills, S., Hobson, R. (2004). Microbiological water quality in a large in-building water recycling facility. *Water Sci. Technol.*, 50(2), 165–172.

Bitton, G. (2005). *Wastewater Microbiology*. John Wiley & Sons, New York.

Blumenthal, U.J., Mara, D.D., Peasey, A., Ruiz-Palacios, G., Scott, R. (2000). Guidelines for the microbiological quality of treated wastewater used in agriculture: Recommendations for revising WHO guidelines. *Bull. World Health Organ.*, 78(9), 1104–1116.

Braile, P.M. and Cavalcanti, J.E.W.A. (1979). *Manual de tratamento de águas residuárias industriais*. CETESB, São Paulo.

Branco, S.M. and Rocha, A.A. (1979), *Poluiçao, proteçao e usos multiplos de represas*, Ed. Edgard Blücher, Sao Paolo.

Chauret, C., Springthorpe, S., Sattar, S. (1999). Fate of *Cryptosporidium* oocysts, *Giardia* cysts, and microbial indicators during wastewater treatment and anaerobic sludge digestion. *Can. J. Microbiol.*, 45, 257–262.

Chu, W.H., Wang, J.Y., Kao, C.M. (2003). A simplified risk-based approach for process screening in municipal wastewater reclamation and reuse. *Water Sci. Technol.*, 47(1), 257–262.

Cloete, T.E., Rose, J., Nel, L.H., Ford, T. (eds) (2004). *Microbial Waterborne Pathogens*. IWA Publishing, London.

Cooper, R.C. and Olivieri, A.W. (1998). Infectious disease concerns in wastewater reuse. In *Wastewater Reclamation and Reuse*, Asano, T. (ed.). CRC Press, New York, 489–520.

Crook, J. (1998). Water reclamation and reuse criteria. In *Wastewater Reclamation and Reuse*, Asano, T. (ed.). CRC Press, New York, 627–703.

Devaux, I., Gerbaud, L., Planchon, C., Bontoux, J., Glanddier, P.Y. (2001). Infectious risk associated with wastewater reuse: An epidemiological approach applied to the case of Clermont-Ferrand, France. *Water Sci. Technol.*, 43(12), 53–60.

Doran, J.W., Alexander, M. (1977). Microbial formation of volatile selenium compounds in soil. *Soil Sci. Soc. Am. J.*, 41, 69–73.

Emerson, K., Russo, R.C., Lund, R.E., Thurston, R.V. (1975). Aqueous ammonia equilibrium calculations: Effect of pH and temperature. *J. Fish. Res. Board Can.*, 32(12), 2379–2383.

Garcia, A., Yanko, G., Widmer, G. (2002). *Giardia* cysts in tertiary-treated wastewater effluents: Are they infective? *Water Environ. Res.*, 74, 541–544.

Gardner, E.A. (2003). Some examples of water recycling in Australian urban environments: A step towards environmental sustainability. *Water Sci. Technol. Water Supply*, 3(4), 21–31.

Gennaccaro, A., McLaughlin, M.R., Quintero-Betancourt, W., Huffman, D., Rose, J.B. (2003). Infectious *Cryptosporidium parvum* oocysts in final reclaimed effluent. *Appl. Environ. Microbiol.*, 69(8), 4983–4984.

Gerba, C.P. and Rose, J.B. (2003). International guidelines for water recycling: Microbiological considerations. *Water Sci. Technol. Water Supply*, 3(4), 311–316.

Government of Canada (2021). Recommandations au sujet de la qualité des eaux utilisées à des fins récréatives au Canada : document de synthèse. Santé Canada, Ottawa.

Haas, C.N. (2000). Epidemiology, microbiology, and risk assessment of waterborne pathogens including *Cryptosporidium*. *J. Food Prot.*, 63(6), 827–831.

Howard, G., Pedley, S., Tibatemwa, S. (2000). Quantitative microbial risk assessment to estimate health risks attributable to water supply: Can the technique be applied in developing countries with limited data? *J. Water Health*, 4(1), 49–65.

IAWQ (1995). Activated sludge model. Scientific and Technical Report no. 2.

Kolpin, D.W., Furlong, E.T., Meyer, M.T., Thurman, E.M., Zaugg, S.D., Barber, L.B., Buxton, H.B. (2002). Pharmaceuticals, hormones, and other organic wastewater contaminants in U.S. streams, 1999–2000: A national reconnaissance. *Environ. Sci. Technol.*, 36(6), 1202–1211.

Krikelis, V., Spyrou, N., Makoulatos, P., Serie, C. (1984). Seasonal distribution of enteroviruses and adenoviruses in domestic sewage. *Can. J. Microbiol.*, 31, 24–25.

Law, I.B. (1996) Rouse Hill – Australia's first full scale domestic non-potable reuse application. *Water Sci. Technol.*, 33(10/11), 71–78.

Lazarova, V., Hills, S., Birks, R. (2003). Using recycled water for non-potable, urban uses: A review with particular reference to toilet flushing. *Water Sci. Technol. Water Supply*, 3(4), 69–77.

Lopez, B. and Laurent, B. (2013). Campagne exceptionnelle d'analyses des substances présentes dans les eaux souterraines. Report BRGM/RP 61853FR, ONEMA – BRGM, Paris.

Lopman, B.A., Reacher, M.H., van Duijnhoven, Y., Hanon, F.-X., Brown, D., Koopmans, M. (2003). Viral gastroenteritis outbreaks in Europe, 1995–2000. *Emerg. Infect. Dis.*, 9(1), 90–96.

Marais, G.R. and Ekama, G.A. (1976). The activated sludge process. Part 1 – Steadystate behaviour, *Water S.A.*, 2(4), 164–200.

Marsalek, J., Schaefer, K., Exall, K., Brannen, L., Aidun, B. (2002). *Water Reuse and Recycling*. Canadian Council of Ministers of the Environment, Winnipeg, Manitoba.

Maunula, L., Miettinen, I., von Bonsdorff, C.-H. (2005). Norovirus outbreaks from drinking water. *Emerg. Infect. Dis.*, 11(11), 1715–1721.

Michael-Kordatou, I., Michael, C., Duan, X., He, X., Dionysiou, D.D. (2015). Dissolved effluent organic matter: Characteristics and potential implications in wastewater treatment and reuse applications. *Water Res.*, 77, 213–248.

Olivieri, A.W., Eisenberg, D.M., Soller, J.A., Suffet, I.H., Froines, J., Cooper, R.C., Gagliardo, P.F. (1998). Comparative chemical risk assessment of repurified water from an advanced water treatment facility and a raw drinking water supply. In *Water Reuse 98, Conference Proceedings*, 135–147.

Ottosson, J. (2002). Faecal contamination of greywater – Assessing the treatment required for hygienically safe reuse or discharge. In *2nd International Symposium on Ecological Sanitation*, Lübeck, 7–11 April.

Ottosson, J. and Stenstrom, T.A. (2003). Faecal contamination of greywater and associated microbial risks. *Water Res.*, 37, 645–655.

Payment, P., Plante, R., Cejka, P. (2001). Removal of indicator bacteria, human enteric viruses, *Giardia* cysts, and *Cryptosporidium* oocysts at a large wastewater primary treatment facility. *Can. J. Microbiol.*, 47, 188–193.

Pessöa, C.A. and Jordao, E.P. (1982). *Tratamento de esgotos domesticos*. ABES, São Paulo.

Rose, J.B., Haas, C.N., Regli, S. (1991). Risk assessment and control of waterborne giardiasis. *Am. J. Public Health*, 81(6), 709–713.

Rose, J.B., Dickson, L.J., Farrah, S.R., Carnahan, R.P. (1996). Removal of pathogenic and indica-tor microorganisms by a full-scale water reclamation facility. *Water Res.*, 30(11), 2785–2797.

Rose, J.B., Farrah, S., Harwood, V.J., Levine, A.D., Lukasik, J., Menendez, J., Scott, T.M. (2004). *Reduction of Pathogens, Indicator Bacteria and Alternative Indicators by Wastewater Treatment and Reclamation Processes.* Water Environment Research Foundation/International Water Association, Alexandria.

Schaub, S.A. and Oshiro, R.K. (2000). Public health concerns about caliciviruses as waterborne contaminants. *J. Infect. Dis.*, 181(Suppl. 2), S374–S380.

Scott, T.T., McLaughlin, M.R., Harwood, V.J., Chicukula, V., Levine, A., Gennaccaro, A., Lucasik, J., Farrah, S.R., Rose, J.B. (2003). Reduction of pathogens, indicator bacteria and alternative indicators by wastewater treatment and reclamation processes. *Water Sci. Technol. Water Supply*, 3(4), 247–252.

Storey, M.V. and Ashbolt, N.J. (2003). A risk model for enteric virus accumulation and release from recycled water distribution pipe biofilms. *Water Supply*, 3(3), 93–100.

Thomann, R. and Mueller, J.A. (1987). *Principles of Surface Water Quality Modeling and Control*. Harper and Row, London.

Toze, S. (2004). Literature review on the fate of viruses and other pathogens and health risks in non-potable reuse of stormwater and reclaimed water. Australian Water Conservation and Reuse Research Program, Commonwealth Scientific and Industrial Research Organisation.

Trevett, A.F., Carter, R.C., Westrell, T., Bergstedt, O., Stenstrom, T.A, Ashbolt, N.J. (2003). A theoretical approach to assess microbial risks due to failures in drinking water systems. *Int. J. Environ. Health Res.*, 13(2), 181–197.

Westrell, T., Schonning, C., Stenstrom, T.A., Ashbolt, N.J. (2004). QMRA (quantitative microbial risk assessment) and HACCP (hazard analysis and critical control points) for management of pathogens in wastewater and sewage sludge treatment and reuse. *Water Sci. Technol.*, 50(2), 23–30.

WHO (2006). A compendium of standards for wastewater reuse in the Eastern Mediterranean Region. Recommendations, World Health Organization, Geneva.

Yates, M. and Gerba, C. (1998). Microbial considerations in wastewater reclamation and reuse. In *Wastewater Reclamation and Reuse*, Asano, T. (ed.). CRC Press, New York, 437–487.

York, D.W. and Walker-Coleman, L. (2000). Pathogen standards for reclaimed water. *Water Environ. Technol.*, 12(1), 58–61.

2

Processing Unit Operations

What happens in a wastewater treatment plant is essentially the same as what naturally happens in a lake, a river or any stream. The function of a wastewater treatment plant is to speed up the natural process of self-treatment. Wastewater treatment has been developed and perfected, using biological, physical, chemical and mechanical techniques. As a result, public health and water quality are better protected today when these treatments are installed.

2.1. Background

Wastewater can be defined as the flow of wastewater discharged by residents, businesses, industries, commercial activities and institutions. This flow is conveyed towards treatment plants by a well-designed network of pipes. The term "domestic wastewater" refers to the flows discharged mainly by populations who use water for food preparation, laundry, cleaning and personal hygiene. Industrial/commercial wastewater is discharged by economic, industrial and commercial activities. Typically, 150–300 liters of wastewater are consumed per day and per person connected to the wastewater network.

The overall objectives of water management and wastewater treatment are associated with the removal of pollutants as well as the protection and preservation of natural water resources.

Wastewater treatment plants involve many processes that, combined all together, achieve the desired goals of water quality.

These operations concern:

– suspended solids;

– colloidal matter;

– organic carbon;

– nitrogen;

– phosphorus;

– micropollutants;

– microorganisms.

These processes require the separation and removal of pollutants existing in wastewater, by means of four basic techniques (physical, mechanical, biological and chemical):

– The physical treatment methods include the use of structures designed to contain and control the flow of wastewater.

– Mechanical techniques involve the use of specific equipment to reduce and restrict pollution loads.

– Bacterial action is optimally used for the biological degradation of pollutants. Bacteria play a fundamental role at the performance level that cannot be matched by other techniques.

– Chemical treatment methods improve the efficiency of unit operations and are used to refine the quality of discharges.

Treatment methods include unitary operations and processes, and their integration constitutes the overall processing chain.

The concepts of unit operations and unit processes are frequently used interchangeably, as they can occur simultaneously in the same reclamation process. In general, the following definitions can be adopted (Metcalf and Eddy (2004):

– physical unit operations: treatment methods in which physical forces are predominant (such as, for example, screening, mixing, flocculation, sedimentation, flotation, filtration, membranes);

– chemical unit processes: treatment methods in which the removal or conversion of contaminants is performed by adding chemical products or through chemical reactions (as precipitation, adsorption or disinfection for instance);

– biological unit processes: treatment methods in which the removal of contaminants is achieved by the biological activity (e.g. organic matter removal, nitrification, denitrification).

2.2. Design principles for treatment plants

Depending on the process used, various mechanisms can act separately or simultaneously for pollutant removal. The main mechanisms are listed in Table 2.1.

Table 2.2 outlines the main processes, operations and treatment systems frequently used in urban wastewater treatment, depending on the pollutant to be removed. These methods are used in the liquid phase, which corresponds to the main flow rate of wastewater supplying wastewater treatment plants. On the other hand, other techniques are associated with solid by-products generated during treatment, in particular sludge.

These include, in particular:

– The treatment of suspended solids implies screening and settling operations. For the screening, traveling screens and sand removal equipment are used.

– Settling achieves a separation through sedimentation or flotation of the liquid and solid phases. Its efficiency is variable according to the particle size and operating conditions. If, for special reasons, biological purification is not possible all year round (if we bear in mind the tourist season, for example), coagulation–flocculation–settling assisted by the addition of chemical reagents can be resorted to. The coagulants will make it possible to neutralize the electrical layer of the colloidal particles, the flocculants will accelerate their agglomeration into flocs and settling will enable the liquid–solid separation.

– Biological treatments used for the treatment of carbon pollution also play an interesting role in eliminating colloidal materials relying on bioflocculation phenomena.

– The treatment of carbon pollution is based on the ability of many bacterial species to degrade organic matter to meet their metabolic requirements. There are three types of technologies: free bacteria (activated sludge, lagoons, oxidation ponds); fixed bacteria (trickling filters, biofilters, biological discs); and hybrid systems with a combination of free bacteria and fixed bacteria on a moving support.

– Nitrogen pollution treatment results in removing ammonia (nitrification processes) and nitrates (denitrification processes). The technologies used are free bacterial systems, attached bacteria or hybrid systems.

– The treatment of phosphorus pollution is based on both biological and chemical processes. It often happens that these two processes are installed in the same wastewater treatment plant, with phosphate precipitation achieved during primary settling, in the activated sludge pond or during sedimentation. Phosphorus removal cannot be applied to biofilters.

– Micropollutant treatment (pesticides, endocrine disruptors, drug residues, etc.) is partly possible during biological treatment, but since not all molecules are blocked, an additional treatment is executed by adsorption (activated carbon), chemical oxidation (ozone) or advanced oxidation (O_3 – UV, O_3 – H_2O_2 – UV), nanofiltration membrane or low-pressure reverse osmosis.

– Microorganism removal is applied when wastewater reuse is scheduled. In this case, a tertiary treatment comprising sandy or bilayer filtration followed by chemical disinfection (Cl_2, ClO^-, ClO_2, O_3, UV) is installed.

Pollutant	**Parameter**	**Treatment technology**
Solids	Dimensions > 1 cm	Screening
	Dimensions > 1 μm	Sedimentation
Organic substances	Particulate BOD	Primary settling
	Dissolved BOD	Sedimentation Physico-chemical treatment Biological treatment
Nitrogen	Ammonia	Biological nitrification
	Nitrite-nitrates	Biological denitrification
Phosphorus	Phosphates	Chemical precipitation Biological phosphate removal
Toxic substances	Pesticides	Adsorption
	Industrial residues	Advanced chemical oxidation
Pathogens	Bacteria	Particle sedimentation Filtration Disinfection
	Viruses	Particle sedimentation Disinfection
	Helminth eggs	Sedimentation Filtration
	Protozoa	Particle sedimentation Filtration Disinfection

Table 2.1. *Main treatment techniques used in wastewater*

The system treatment diagrams described in Tables 2.1 and 2.2 are presented in Figures 2.1–2.8.

Pollutant	Treatment technology
Solids	Screening Sand removal flotation Sedimentation
Organic substances	Free bacteria (activated sludge and derivatives) Fixed bacteria (biofilters and derivatives) Hybrid systems (free biology and fixed bacteria on support) Lagoons and oxidation ponds Biological discs Membranes associated with activated sludge Anaerobic biological reactors Physico-chemical treatment
Nitrogen	Biological nitrification (free and attached bacteria) Biological denitrification (free bacteria and fixed bacteria)
Phosphorus	Chemical precipitation Biological phosphate removal
Toxic substances	Adsorption Advanced chemical oxidation Nanofiltration or low-pressure reverse osmosis membranes
Pathogens	Particle sedimentation Oxidation ponds Filtration Chemical disinfection Ultraviolet radiation disinfection, membranes

Table 2.2. *Treatment operations frequently used for urban wastewater reclamation*

Municipal wastewater treatment therefore comprises a wide range of physical, chemical and biological processes whose purpose is to remove the pollutant load and achieve a final product that can be released into receiving environments or reused after further treatment.

Among the several processes involved, physical processes, which are also commonly referred to as unit operations, play a major role throughout the various treatment stages.

2.3. Unit operations

In a wastewater treatment plant (WWT), an example of which is shown schematically in Figure 2.1, typical steps for good efficiency in terms of biochemical oxygen demand (BOD) and suspended solids can be easily identified:

1) Preliminary treatment (also called pretreatment): aims at removing bulky solids, thus preconditioning the effluent for further treatment. This step involves several unit operations such as screening, grinding, sedimentation for sand removal, skimming and flow equalization.

2) Primary treatment: unit operations such as sedimentation and flotation are used to remove suspended and colloidal fractions from the effluent.

3) Secondary treatment: aims at eliminating organic matter biologically.

4) Complementary treatment: aims at removing nutrients such as nitrogen and phosphorus, suspended residual matter, inorganic and toxic matter, and refractory organic matter that has escaped from the previous steps. This can be done following physical (such as filtration, screening, air stripping, adsorption, ion exchange), chemical (precipitation, oxidation) or biological methods. However, with the rare exception of the very few countries that have resourceful disposal regulations for ecological reasons, typical WWTs do not operate beyond secondary treatment.

5) Disinfection: removes pathogens from water by chemical means (for example, chlorination or ozonation) or physical processes (UV irradiation, for example). The treated effluent can then be safely discharged into natural receiving environments or partially reused (for instance, for irrigation).

6) Sludge treatment: this includes processes such as thickening, dewatering, drying and digestion for the purpose of reducing the volume of sludge to be treated along the processing chain as well as to biologically stabilize the final product, which is usually then sent to discharges (although it can also be used as a fertilizer or fuel).

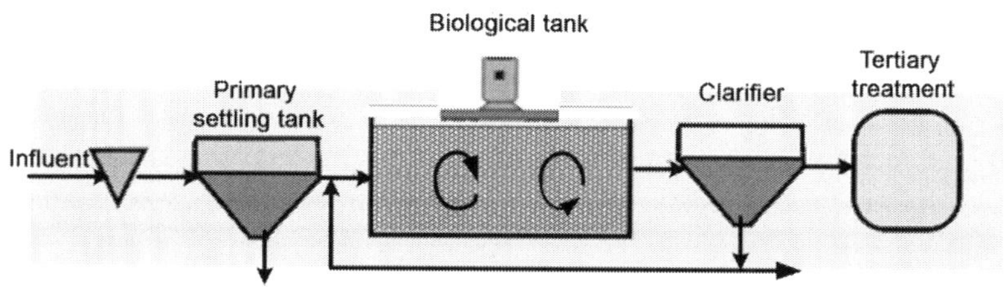

Figure 2.1. *Activated sludge biological treatment process*

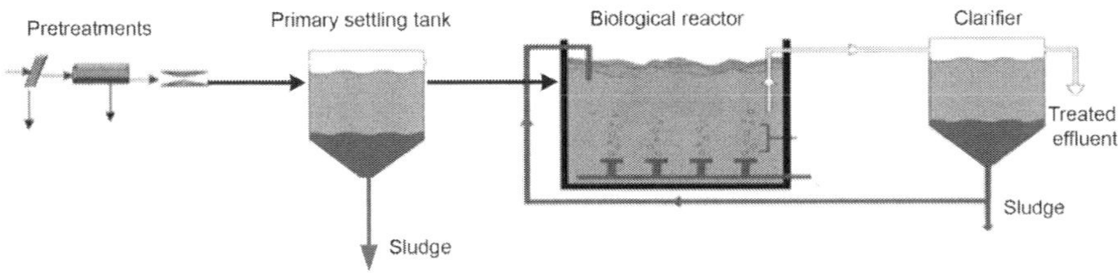

Figure 2.2. *Activated sludge process*

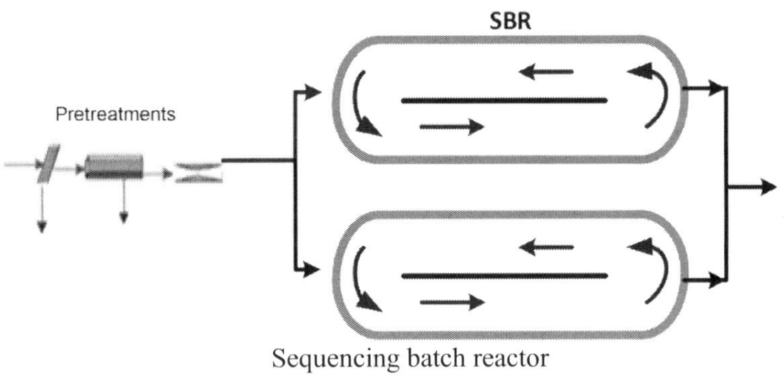

Figure 2.3. *Sequencing batch reactor (SBR) process*

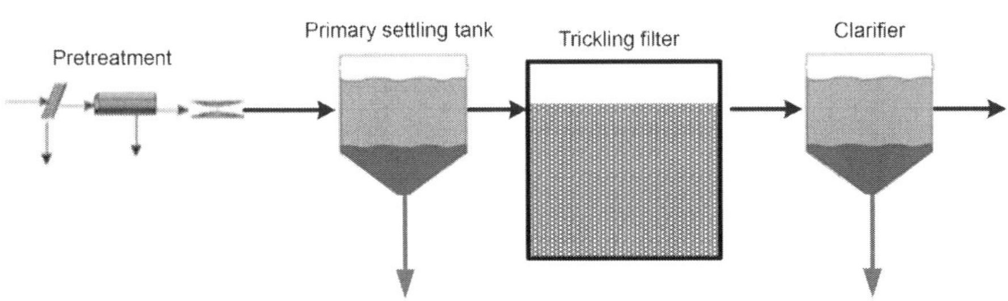

Figure 2.4. *Trickling filter-based process*

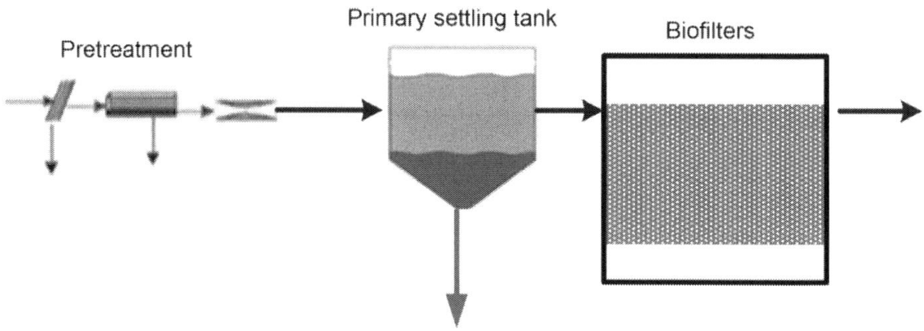

Figure 2.5. *Biofilter-based process*

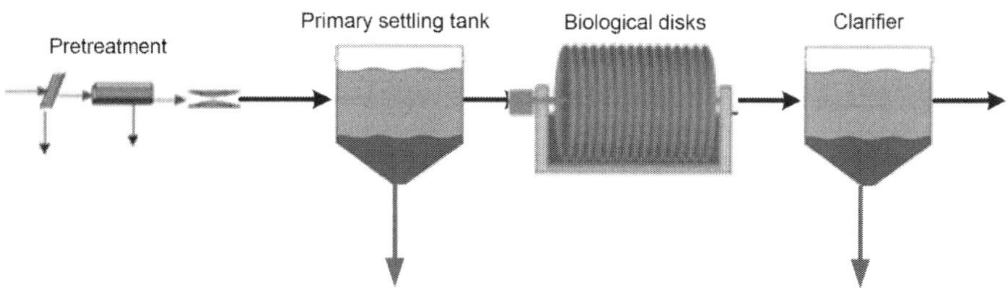

Figure 2.6. *Biological disc-based process*

The purpose of this chapter is to describe the unit operations of wastewater treatment plants.

A wastewater treatment plant is a facility whose purpose is to purify domestic wastewater, industrial water and rainwater before discharging them into the natural environment, usually a watercourse.

To avoid polluting the natural environment, wastewater is cleaned of its many impurities: solids, excrement, oils and all dissolved substances. Their high concentration would make it impossible to directly discharge into the natural environment without generating an aggravated pollution.

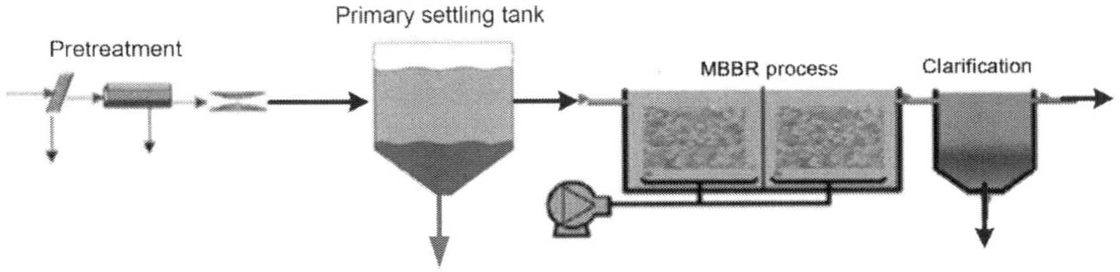

Figure 2.7. *Activated sludge-MBBR hybrid process*

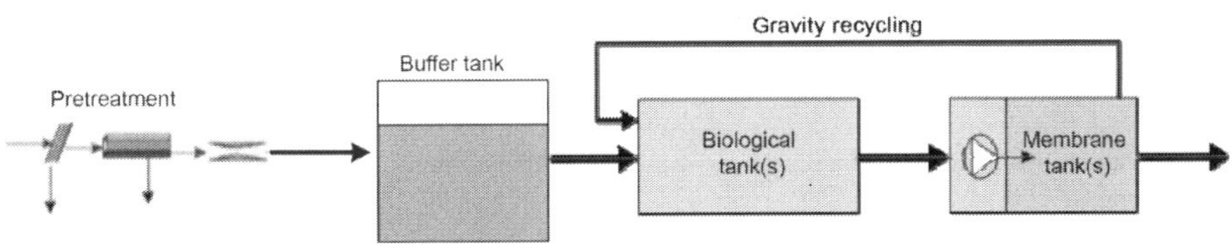

Figure 2.8. *Membrane process (MBR)*

A wastewater treatment plant is generally located at the end of a collection network. It will use various physical and biochemical processes and devices to decompose organic matter and separate it from water.

The elimination of carbon pollution (as well as nitrogen and phosphorus) present in wastewater of mainly domestic origin is conventionally carried out using an aerobic biological process in the presence of biomass in solution (suspended growth or activated sludge) or support-attached biomass (fixed-film).

2.3.1. *Pretreatment*

The aim of pretreatments is to remove the coarsest elements. These are bulky waste (screening), sand and gravel, and grease (degreasing, de-oiling).

During *screening*, wastewater travels through a grate in which bars retain the largest materials. Sieving can supplement this pretreatment phase.

Sand and gravel can be removed from wastewater by sedimentation. These particles are then sucked up by a pump. The recovered sands are drained, then washed before being either discharged or reused, depending on the washing quality.

Degreasing aims to eliminate the presence of grease in wastewater, grease that can interfere with the effectiveness of the biological treatments that take place afterwards. Degreasing is carried out by flotation. The injection of air at the bottom of the system enables fat substances to rise up to the surface. The grease is scraped off the surface and then disposed of (landfill or incineration).

2.3.2. *Primary treatment*

Primary treatment resorts to physical processes, with more or less successful settling, possibly accompanied by physico-chemical processes, such as coagulation–flocculation.

Conventional *primary settling* consists of a separation of liquid and solid elements under the effect of gravity. Solids are deposited at the bottom of a vessel called a “settler” to form the “primary sludge”. This treatment removes 50–55% of suspended solids and reduces BOD and COD by approximately 30%.

The use of a lamellar *settler* results in increasing the efficiency of the settling. This type of vessel comprises inclined parallel lamellae, which multiplies the settling surface and accelerates the particle deposition process. Lamellar settling enables us to remove more than 70% of suspended solids and decreases COD and BOD by more than 40%.

Settling proves even more effective when accompanied by prior flocculation. *Coagulation–flocculation* enables removing up to 90% of suspended solids and 75% of BOD. This technique involves a first phase in which a reagent is added, which causes agglomeration of the suspended particles. The masses of solids obtained in this way are called “flocs”.

2.3.3. *Secondary treatment*

2.3.3.1. *Biological treatment*

In most cases, carbon and nitrogen pollution removal is based on biological processes. The biological treatments reproduce the phenomena of self-purification that exist in nature. Self-purification includes all of the processes by which an aquatic environment manages to regain its original quality after pollution.

Biological purification techniques use the bacterial activity present in the wastewater, which decomposes organic matter. Actually, the “activated sludge” process is mostly widespread in wastewater treatment plants providing secondary treatment.

Among biological treatments, we can find extensive and intensive biological processes.

2.3.3.2. *Extensive biological processes*

Lagoons use the purifying capacity of natural water reservoirs. Wastewater is directed through a series of basins. Oxygen is brought in by exchanges with the atmosphere. Organic pollution degrades under the action of bacteria present in the natural water reservoir. This purification method makes it possible to remove 80–90% of the BOD and 20–30% of the nitrogen and contributes to a very significant germ reduction. It has nonetheless the disadvantage of using large areas.

2.3.3.3. *Intensive biological processes*

They include a whole series of techniques that share in common the use of bacterial cultures that "consume" polluting materials. There are two main categories of artificial biological processes:

– *"Activated sludge" facilities*: these are aerobic purification systems, which require supplying oxygen. The bacterial culture is maintained in an aerated mixing tank. The organic matter in the water is converted into carbon (in the form of carbon dioxide, CO_2) through the action of bacteria. The excess of bacteria thus formed are called "sludge". After a period of residence in an aerated lagoon, the effluent is redirected to a clarifier, also called a secondary settler. The sludge is then either sent to a specific treatment unit for agricultural spreading or disposal, or partly reinjected into the aerated lagoon. This operation is referred to as "sludge recirculation". Activated sludge-based treatments remove 85–95% of BOD_5, depending on the facility. It is the simplest and most frequent biological treatment used.

– *"Fixed-film" facilities*: the trickling filter technique consists of running off the water to be treated over a solid support where a culture of purifying microorganisms develops, named "biological film" or "biofilm". The maximum efficiency of this technique is 80% BOD removal. They are generally preferred for facilities with a size of less than 2000–5000 population equivalents.

Biofiltration uses a bacterial culture attached to a granular support. The granular medium serves as both a filter and a support for bacterial cultures. This process therefore makes it possible to both decompose pollutants and clarify wastewater. Regardless of its nature, the selected material must be characterized by its filtering action and allow maximum fixation of biological cultures. An aerated system supplies the necessary oxygen inside the filter. This technique removes about 90% of BOD and can also remove nitrogen.

2.3.3.4. *Membrane processes*

Membrane processes combine biological and physical processes. These very thin membranes act as a physical barrier that blocks microorganisms and particles.

2.3.4. *Effluent clarification and discharge*

At the end of the treatment, a final settling allows the separation of purified water from sludge resulting from the growth after the degradation of organic matter. This settling is achieved in special tanks known as *clarifiers*. The purified water can then be discharged into the natural environment.

2.3.5. *Additional treatments*

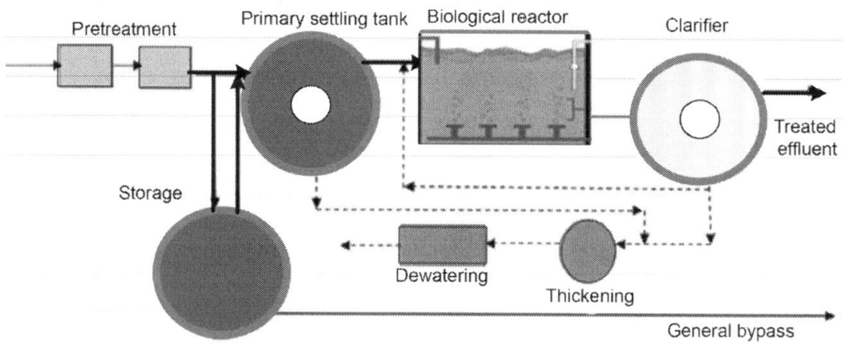

Figure 2.9. *Example of a biological wastewater treatment plant*

Additional treatments concern nitrogen and phosphorus removal in order to comply with increasingly stricter environmental standards. In addition, further unit operations can enhance the treatment process, such as disinfection, micropollutant removal, sludge recovery, etc.

2.3.5.1. *Nitrogen removal*

Wastewater treatment plants designed to eliminate carbonaceous materials remove only about 20% of the nitrogen present in wastewater. To comply with discharge standards in sensitive areas, additional treatments must be put in place. Nitrogen and phosphorus are likely to lead to water eutrophication phenomena, in the form of algal proliferation leading to fish mortalities. This phenomenon mainly concerns areas sensitive to eutrophication in which wastewater treatment plants must be equipped with processes allowing nitrogen and/or phosphorus removal.

However, the purification of the nitrogen and phosphorus components and their transformations in the wastewater treatment plants will greatly differ depending on whether nitrogen or phosphorus is being treated.

Under the action of bacteria, nitrogen will be submitted to a series of oxidations, then successive reductions bringing it from the NH_4^+ stage (ammonium ion, oxidation state +1), to the NO_2^- stage (nitrite, oxidation state +3), then nitrate (stage +5) and then reduced to the N_2 stage (nitrogen: zero oxidation state). At this stage, it is in a gaseous form and can then travel into the atmosphere where it will join atmospheric nitrogen. Nitrogen is therefore eliminated in gaseous form.

The necessary conditions for the development of bacteria responsible for the oxidation stages are not very difficult to achieve in a WWTP. However, these microorganisms consume a lot of oxygen (oxidation reactions) that must be supplied by mechanical means (turbines, blowers), which implies energy consumption and therefore operating costs.

The reduction (denitrification) steps do not require oxygen, but instead allow some of the oxygen that was originally used to be recovered. These operations nevertheless require special implementations that cannot be carried out in old or overloaded WWTP

Organic nitrogen is converted into ammonia nitrogen (NH_4^+) in wastewater. The removal of ammoniacal nitrogen is most often achieved by means of biological treatments, such as *nitrification–denitrification*. Nitrification consists of the transformation, by bacterial cultures, of ammoniacal nitrogen into nitrates (NO_3^-), an oxidized form of nitrogen. A second phase of denitrification completes the process. Under the action of "denitrifying" bacteria, nitrates are transformed into nitrogen gas. This gas then escapes into the atmosphere.

2.3.5.2. *Phosphorus removal*

Phosphorus removal or *dephosphatation* can be carried out by physico-chemical or biological processes.

Phosphorus can be extracted from the liquid phase and discharged in solid form. In fact, there are no gas forms of this element in the phosphorus transformations implemented in a wastewater treatment plant. It will therefore have to be extracted in the form of solids: either in the form of salts obtained by chemical precipitation from appropriate reagents, or as an integral part of the biological sludge produced in the treatment plant.

In *physico-chemical treatment*, the addition of reagents, such as iron or aluminum salts, yields a precipitation of insoluble phosphates and their removal by settling. These techniques, which are currently the most widely used, remove between 80 and 90% of phosphorus, but generate significant sludge production.

The *biological removal* of phosphorus corresponds to the second case. The purpose is to obtain a biological sludge with a particularly high phosphorus content. This unusual enrichment is only achieved by alternating the environmental conditions of the bacteria responsible for this phenomenon (creation of a state of stress for bacteria).

Biological phosphate removal consists of promoting phosphorus accumulation in bacterial cultures. The average yield is approximately 60%.

In the first case, the operational cost will be linked both to the cost of the reagents involved and to the investments for the equipment for the recovery of the formed precipitates.

If the biological phosphorus removal operation is in itself inexpensive in terms of energy, it requires more expensive investments at the beginning and requires a much more accurate management of the plant than a conventional WWTP.

It should be noted that in both cases (physico-chemical or biological treatments), phosphorus will be found in the sludge produced by the plant. It will then be associated with the fate of the sludge and may therefore, depending on the case, be used for incineration or direct agricultural recovery.

Phosphorus is a useful and valuable element whose resources are not infinite. It is increasingly considered that it should be extracted from the sludge for its recovery. It is currently possible to induce the precipitation of a particular compound, struvite, which is a nitrogen and phosphorus salt, which can be used as a fertilizer containing both the nitrogen and phosphorus needed for crops. This then allows for the recovery and recycling of these two essential elements for agriculture.

2.3.5.3. *Disinfection*

Primary and secondary treatments do not completely destroy the germs present in urban wastewaters. Additional disposal processes are therefore used when treated water is reused for different applications.

Chlorine is the most common disinfectant. Disinfection can, however, also be performed with ozone or bromine, chlorine dioxide, chloramines, peracetic acid, performic acid and ultraviolet systems.

2.3.5.4. *Micropollutant removal*

To date, organic carbonaceous pollution and its removal have been monitored by means of global measurements such as COD (chemical oxygen demand) and BOD (biochemical oxygen demand). These two global parameters are oxygen demands, that is, the amount of oxygen that is required to degrade pollution is measured, either chemically (COD) or biologically (BOD). These are important parameters because in treatment plants, the oxygen required must be artificially supplied and the operational cost will be proportional to the measured values of oxygen demand (BOD, COD).

In contrast, domestic or industrial wastewater contains a large variety of different pollutants and BOD (or COD) measurements do not provide any indication of the content of these individual compounds. Yet not all these products are degraded with the same efficiency when they travel through treatment plants.

Therefore, in the outlet of the wastewater treatment plant, we will find low concentrations of non-degraded products, partially degraded products and transformation residues. Indeed, this is somewhat well-established in the minds of our fellow citizens and yet it is false concerning the possibilities of complete degradation by bacteria with regard to organic micropollutants. It is usually thought that, if given the time, microorganisms will always end up totally decomposing the substances which they are in contact with.

However, this is certainly not true, at least on the time scale of a few months or a few years.

We should bear in mind natural substances such as humic acids found in forest soils. Despite being natural, these substances are very poorly degraded by microorganisms; otherwise, they would be oxidized into CO_2 and would not accumulate in the soil.

It is true that many substances of natural origin are biodegradable; in other words, they can be broken down by microorganisms.

This property is not as frequent for synthetic organic products. Therefore, a rather large number of organic molecules produced by chemical synthesis are in fact not very biodegradable or not at all. They can then travel through treatment plants without undergoing any major transformations, and finally end up in the receiving environments. Due to the fact that decades-long sanitation policies and advances in measurement technology have allowed for the recovery of the quality of watercourses, it is now possible to detect these pollutants in trace states.

This pollution was previously occulted by primary (SS), secondary (BOD, COD) or tertiary pollution (N, P); since these latter have been subjected to purification efforts, there is now increased attention to micropollutants, which some authors also classify as "quaternary pollution".

Consequently, their presence in receiving environments, their rate of reduction in conventional wastewater treatment plants and even the types of "quaternary" treatments that will have to be implemented to further degrade them are topics that receive most attention.

Among these substances, we should of course mention the substances included in the list of toxic substances, pesticides, medications or their metabolites, estrogens as well as micropollutants with endocrine modulating effects, steroids, metabolites of detergents and products such as perfumes, etc.

Although most of them have been in use for a few decades now, these substances are considered to be potentially disruptive to ecosystems, not only because of direct toxicity effects that have most often led to their use being banned, but also because they are likely to reduce the fertility of certain species (fish, crustaceans, shellfish, etc.). The issue of the potential risk to human health is now a major concern.

This new environmental issue is the subject of in-depth studies to measure the contamination rates of aquatic environments, quantify the reductions obtained in the various treatment systems and better identify the flows of these materials in plants (do these pollutants mainly exist in treated water, in sludge or in fatty products (fats)?) and finally, better identify the risks associated with their presence, eventually in a simultaneous manner, in these ecosystems.

2.3.6. *Sludge treatment*

2.3.6.1. *Volume reduction*

Regardless of the ultimate destination of sludge, it is imperative to reduce its volume, that is, its water content. This reduction reduces storage and transport costs for the purposes of recycling and gives homogeneity: the reduction facilitates handling and increases the calorific value in case of incineration.

The processes for volume reduction are as follows:

– *Thickening*: this is the first step in the treatment. Thickening is carried out by gravity settling and flotation (flotation consists of dissolving air in the incoming sludge inside a tank. During their ascent, the bubbles carry the suspended solids to the surface, by draining – gravity-driven filtration with polymer supply).

– *Dewatering*: this results in the reduction of the volume of sludge by the concentration effect. It is operated by centrifugation (the water is separated from the sludge under the effect of acceleration) or by draining systems with the addition of polymer.

– *Drying*: this consists of evaporating the water contained in a sludge that has already been dewatered. Drying is more or less extensive (30 or 45% dry content partial drying or 60–90% dry content advanced drying) depending on subsequent use. Extensive drying leads to increased agricultural recovery and, in the case of incineration, guarantees a calorific value comparable to that of domestic waste.

2.3.6.2. *Reclamation*

2.3.6.2.1. Stabilization and sanitation

A high load of volatile matter remains in the sludge as organic matter that tends to ferment, especially in the absence of air, and to generate olfactory nuisances. It may also contain undesirable pathogens.

The objective of stabilization is to prevent sludge fermentation after treatment. It also indirectly results in reducing odors. Stabilization can be achieved by biological or physical means, and specially by heating. A sludge with a high amount of dry matter contains very little water, which prevents any microbial growth.

Sanitation is the removal of pathogens. Sanitation requires temperature conditions for a given time and/or a pH that is either very high (the increase in pH is ensured by liming, that is, by adding lime, which allows pH to be increased to 12) or very low (by adding nitrite, which reduces pH to 2).

Stabilization and sanitation can be performed by biological means (composting), chemical means such as lime addition, or thermal systems (drying). The differences between treatments are notable in the duration of the treatment and temperatures.

2.3.6.2.2. Composting

Composting consists of making a mixture of fresh sludge and co-products that will form carbonaceous supports that will be capable of absorbing the water from the sludge (wood, green waste, etc.), and then letting the whole thing develop for several weeks, during which fermentation takes place. The procedure creates compost. Aeration is provided during this process. It can be carried out by simple overturning or it can be sped up when fermentation takes place in a closed silo. Composting makes it possible to dry sludge to up to 60% dry solids.

The agronomic value of compost lies in its fertilizer content (essentially nitrogen) and humic matter. Maintaining the humus level promotes water retention, reducing erosion and gullying. Spreading compost ensures good revegetation in soils depleted by mineral fertilizers.

The composting technique is experiencing an important development. There are various products serving as support:

– composting with wood residues;

– composting with green waste and pallets;

– composting with wood waste and addition of consumer oils;

– composting with domestic wastes.

Sludge, such as animal excrement, contains living microorganisms and contaminants.

Risks related to pathogen microorganisms are short-term risks related to their very short lifespan in the soil, which varies from a few days to a few months. The origin of these microorganisms is directly linked to septic processes linked to the degradation of organic matter in collection networks and then in the treatment plant. It is possible to carry out a sanitation treatment to reduce or eliminate them through digestion, thermophilic stabilization, composting and liming. However, sewage sludge is not a favorable environment for the survival of these pathogenic microorganisms.

2.3.6.2.3. Spreading

In addition, spreading accelerates sludge destruction by subjecting it to climatic hazards (temperature, solar radiation) and soil effects. Soil interactions relate to competition with other microorganisms and soil microflora, the amount or absence of organic matter, pH level and soil composition. Furthermore, the health risk for humans is minimal if the regulatory precautions for use are respected. These precautions are improved if the spreading is carried out in grasslands.

The environmental risks associated with contaminants present in sludge include those related to trace organic compounds from detergents, solvents and household pesticides, etc., and trace elements of heavy metals, the risks of which are linked to the accumulation mechanisms in soils.

The notion of danger needs to be put into perspective. Soils naturally contain high levels of trace elements (heavy metals); they represent a cumulative surface area that is significantly larger than that reserved for sludge spreading and are not a metal-enhancing medium as marine chains.

In order to avoid these minimal risks, French regulations set maximum concentrations and maximum annual fluxes of trace elements that must not be exceeded. Compliance with these standards is enough for protecting against the risks caused by the presence of the undesirable elements previously mentioned.

The conditions for spreading defined by the decree cover five operations:

– storage;

– studies prior to spreading;

– the provisional spreading program, which consists of planning the spreading periods;

– a spreading register that must be kept by the farmer;

– an annual report.

2.3.6.3. *Sludge thermal treatment*

Wet oxidation (WO) consists of heating a sludge at high temperature and under pressure with an oxidizing gas (air or oxygen) so as to transform some of its organic compounds by oxidation.

Sludge contains three types of components: water in varying proportions depending on the degree of dry solid content and dry matter consisting of mineral and organic matter in the sludge. During conventional thermal treatment, on dried sludge, the organic matter contained in the sludge is oxidized and the sludge is destroyed at a very high temperature (900°C). Wet treatment makes it possible to remove simply pasty sludge by heating it at high temperature (250–300°C) but above all under high pressure to destroy the organic matter. The pressure rise is a result of the fact that water turns into steam below a certain temperature and that pressure must be increased to keep the water in a liquid state. The treatment yields water, gas and a mineral residue.

A new generation of this system, marketed by Veolia and named the ATHOS® process, is capable of reducing temperature (235°C) and pressure with a supply of pure oxygen. Organic matter is no longer destroyed. Part is transformed into gas (CO_2) and part into water-soluble form (acetic acid, nitrogen transformation, sulfur and chlorine into ammonia, in sulfates and water-soluble chlorides). Dewatering involves mineral matter only.

After treatment, the sludge is transformed into three compartments: water with all the compounds that have been dissolved and which can be removed by biological water treatments, gas (CO_2 and nitrogen) and an inert “cake” made up of minerals, aluminum, phosphates, carbonates and heavy metals. These elements have been crystallized and are therefore found in inert form. In France, they can go to a class III landfill or be used as a sublayer for road work. These products are patented by the company under the name "technosable".

The interest of this type of treatment is threefold. On the one hand, this technique starts from thickened sludge and not from dried or dewatered sludge, which avoids pretreatment. On the other hand, the final product is extremely reduced: a ton of sludge with 4% dry content, that is, 40 kg of dry solids, gives 10 kg of "technosable". Finally, the company advertises a competitive price, comparable to, or even slightly cheaper than, that of incineration.

2.4. References

Brix, H. (ed.) (1993). Wastewater treatment in constructed wetlands: System design, removal processes, and treatment performance. In *Constructed Wetlands for Water Quality Improvement*. CRC Press, Boca Raton, 9–22.

van Buuren, J.C.L., Frijns, J.A.G., Lettinga, G. (1995). *Wastewater Treatment and Reuse in Developing Countries*. Wageningen Agricultural University, Wageningen.

Campos, H.M. and Von Sperling, M. (1996). Estimation of domestic wastewater characteristics in a developing country based on socio-economical variables. *Water Science and Technology*, 34(3/4), 71–77.

Chagnon, F. and Harleman, D.R. (2002). *An Introduction to Chemically Enhanced Primary Treatment*. Massachusetts Institute of Technology, Cambridge.

Environmental Protection Agency (EPA) (1985). *Rates, Constants, and Kinetics Formulations in Surface Water Quality Modeling*. EPA, Athens.

Environmental Protection Agency (EPA) (1987). *Design Manual. Dewatering Municipal Wastewater Sludges*. EPA, Cincinnati.

Environmental Protection Agency (EPA) (1998). *How Wastewater Treatment Works... The Basics*. EPA, New York.

Gandhi, V. and Shah, K. (2021). *Advances in Wastewater Treatment I*. Materials Research Foundations, Millersville.

Gedda, G., Balakrishna, K., Devi, R.U., Shah, K.J. (2021). Introduction to conventional waste-water treatment technologies: Limitations and recent advances. In *Advances in Wastewater Treatment I*, Gandhi, V. and Shah, K. (eds). Materials Research Foundations, Millersville, 1–36.

Guyer, J.P. (2018). *An Introduction to Preliminary Wastewater Treatment*. Guyer, New York.

Lekang, O.I. and Kleppe, H. (2000). Efficiency of nitrification in trickling filters using different filter media. *Aquacultural Engineering*, 21(3), 181–199.

Marais, G.R. and Ekama, G.A. (1976). The activated sludge process. Part 1 – Steady state behaviour. *Water S.A.*, 2(4), 164–200.

Metcalf & Eddy, Inc. (2004). *Wastewater Engineering: Treatment and Reuse*. McGraw-Hill, New York.

Nathanson, J.A. and Ambulkar, A. (2025). Wastewater treatment. *Encyclopaedia Britannica*.

Pennsylvania Department of Environmental Protection (2016). Module 1: Introduction to wastewater treatment. Revised 2016. Pennsylvania Department of Environmental Protection, Pennsylvania State Association of Township Supervisors (PSATS), Gannett Fleming, Inc., Dering Consulting Group, Penn State Harrisburg Environmental Training Center.

Ramalho, R.S. (1977). *Introduction to Wastewater Treatment Processes*. Academic Press, New York.

Tatoulis, T.I., Tekerlekopoulou, A.G., Akratos, C.S., Pavlou, S., Vayenas, D.V. (2015). Aerobic biological treatment of second cheese whey in suspended and attached growth reactors. *Journal of Chemical Technology and Biotechnology*, 90(11), 2040–2049.

Von Sperling, M. (2007). *Basic Principles of Wastewater Treatment: 2*. International Water Association, New York.

Von Sperling, M. and Chernicharo, C.A.L. (2002). Urban wastewater treatment technologies and the implementation of discharge standards in developing countries. *Urban Water*, 4, 105–114.

Von Sperling, M. and Goncalves, R.F. (2002). Sludge production in wastewater treatment systems in warm-climate countries. In *3rd IWA World Water Congress*. International Water Association, Melbourne, 7–12 April.

3

Estimation of Pollutant Loadings

The population that contributes to the treatment plant is the one located inside the design area supplied by the wastewater collection network. Therefore, the concept of domestic flow concerns wastewater discharged from homes, as well as commercial activities that contribute to the development of the urban area.

3.1. Water consumption

The flow of domestic wastewater is calculated from the overall water consumption in the locality. This consumption is generally calculated on the basis of the reference population and a value assigned for the daily average per person.

Minimum and maximum flows are also estimated in order to avoid the hydraulic and physical constraints that can affect the operation of the WWTP.

	Population (inhabitants)	**Consumption per capita (L/person/d)**
Rural commune	$< 5,000$	90–110
Village	5,000–10,000	100–150
Small city	10,000–50,000	110–170
Medium-sized city	50,000–250,000	120–200
Big city	$> 250,000$	200–300

Table 3.1. *Water consumption according to the number of connected inhabitants*

The levels of water consumption per capita can vary from one place to another and from one country to another (Table 3.1). The daily volume per capita depends on the number of inhabitants connected, which itself is linked to the degree of development of the urban population. It most often reaches 150 L/capita/day, but in many cities, it can vary between 200 and 300 L/person /day.

The factors that affect this parameter are:

– water availability;

– climate: hot countries induce higher water consumption;

– the community size: large cities generate higher water consumption per capita due to strong economic and industrial activity;

– the economic level of the community: a higher level is associated with higher water consumption;

– the level of industrialization: industrialized areas tend to exhibit a higher level of water consumption;

– water cost: a higher cost reduces water consumption;

– water pressure: high pressure in the distribution system leads to greater use and waste of water;

– water losses: losses in the water distribution network imply the need for a greater quantity of water produced.

3.2. Estimated discharges

Water pollution is composed of various substances from various origins. It generally manifests itself in four main forms: it can be of organic, microbiological, toxic (mineral and organic) or finally only mineral origin, through the discharge, for example, of fertilizing products.

The characterization of pollution used in software simulating the operation of biological wastewater plants uses discrimination by molecule size (settleable, non-settleable, dissolved), by chemical distribution (mineral, organic) and by degree of biodegradability (easily, difficult and non-biodegradable).

Each of these forms of pollution necessarily corresponds to a change in the receiving environment.

The protection of the environment inevitably involves the treatment of these discharges.

3.2.1. *Wastewater flows*

Wastewater treatment (collection, treatment and disposal) is achieved by the following technical alternatives:

– off-site remediation;

– separate network system;

– unified network system;

– in situ sanitation.

The separate network collects rainwater and domestic wastewater separately. The water is transported by independent pipe systems.

In this case, rainwater is not involved in the measuring of wastewater flows to the WWTP.

The unified network collects wastewater and rainwater together in the same pipes that convey them to the WWTP.

In this case, the design of the WWTP takes into account the fraction of rainwater both in terms of hydraulic flow and pollution flow.

In hot countries, during the dry season, wastewater flows slowly through these large-diameter pipes, resulting in long retention times that favor decomposition and the generation of bad smells.

Urban wastewater can carry three main sources of pollution to the WWTP:

– domestic wastewater;

– infiltrations;

– industrial effluents (origins and types of various industries).

It may prove necessary that each of these three sources be analyzed separately in order to estimate specific pollutant fluxes and to evaluate the elements that may be detrimental to the proper functioning of the plant.

The definitions of the various wastewater flows and rainwater flows are classified as follows:

– nominal daily flow ($m^3.d^{-1}$): this flow is normally given in the specifications document and should correspond to the 95th percentile of all daily flows of the dry days per year;

– hourly nominal flow ($m^3.h^{-1}$): nominal daily flow/24 (refers to dry weather);

– daily wet weather flow rate ($m^3.d^{-1}$): without explicit indication in the specifications, twice the nominal daily flow rate is used by default;

– average daily flow ($m^3.d^{-1}$): average daily flow recorded over a year. This average includes all dry and wet weather days;

– dry weather peak flow, Qpdw ($m^3.h^{-1}$): Qpdw = dry weather peak flow coefficient × hourly nominal flow;

– peak rainy flow, Qprf ($m^3.h^{-1}$): Qprf = peak rainy time flow coefficient × hourly nominal flow.

3.2.2. *Peak flows*

– Peak concentration coefficient: without indication in the specifications, it is set at 1.5. This is a peak for dry weather.

– Peak flow coefficient for dry weather, also called peak hourly coefficient: without indication in the specifications, it is calculated by the following empirical law:

$$Cp = 1.5 + \frac{2.5}{\sqrt{Qm}}$$

where Qdf = nominal daily flow rate expressed in $l.s^{-1}$.

– Peak rainy flow coefficient: without indication in the specifications, it is set at 3. This coefficient applies only to unified networks.

– Duration of the peak of dry weather pollution: this is the duration during which we simultaneously get the peak concentration of dry weather and the peak flow of dry weather. This coefficient is used to define the peak O_2 requirements. If not indicated in the specifications, it is set at 2 h.

– Separate network: sewer network collecting only domestic water, rainwater does not disturb the flow.

– Combined network: rainwater is mixed with domestic water in the same sewer network.

3.2.3. *Number of population equivalents (PE)*

The volume of water consumed per capita increases with the size of the agglomeration and varies from country to country. In Europe, it varies between 150 and 400 l/PE/day, depending in general on the size of the concerned urban area.

Conventionally, according to the laws in place in Europe, a population equivalent produces between 55 and 70 g $BOD_5.d^{-1}$.

Using the following expression, the number of population equivalents can be defined from the BOD_5 flux and a unit discharge considered at 60 g $BOD_5/PE.d^{-1}$:

Number of population equivalents = daily BOD_5 flux ($kg.d^{-1}$)/0.06

Four types of flows can be identified:

– the average daily hourly flow;

– the average daytime hourly flow;

– the peak dry weather flow;

– the peak rainy weather flow (combined network).

The average hourly daily flow is the average flow observed during the day, measured at the inlet of the treatment plant. In the general case, the recording of the average hourly wastewater flows present an identical pattern illustrated by a bell curve.

There is a very low flow between midnight and 7 a.m., a fast increase in flow between 7 a.m. and 9 a.m., a period of high flow between 9 a.m. and 2 p.m. during which the peak is at the beginning or end of the morning, depending on the case; this is followed by a trough in the afternoon and a recovery at the end of the day until 8 p.m. to 10 p.m. (maximum), followed by constant decline until midnight (see Figure 3.1).

The flow rates of urban wastewater traveling through a sewerage network can be estimated on the basis of about 150 L per day and per PE of domestic water.

The water flow varies according to the day, the month, the season, the activity, etc. It can be represented by the curve depicted in Figure 3.1.

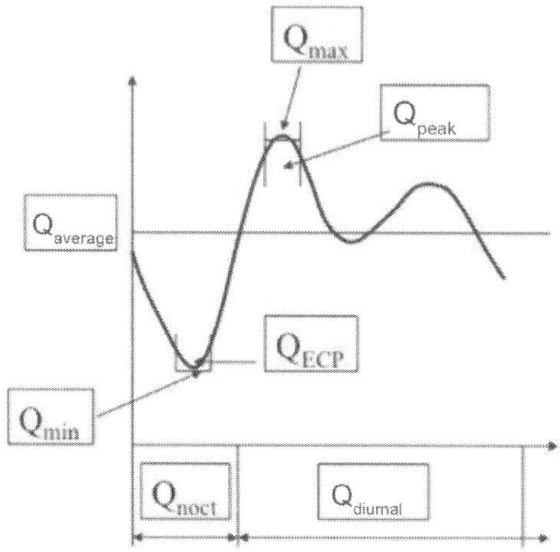

Figure 3.1. *Flow evolution in a day or a month*

COMMENT ON FIGURE 3.1.–

Q_{max}: *maximum flow rate* ($m^3.s^{-1}$).

Q_{min}: *minimum flow rate* ($m^3.h^{-1}$).

Q_{PCW}: *permanent parasitic clear water flow rate with K= 0.7 up to 0.8. The average flow over the night period (8 h) is given by:*

$$Q_{PCW} = Q_{min} \times K\, Q_{night}$$

$Q_{diurnal}$: *average flow over the diurnal period (14 h).*

$Q_{average}$: *average daily flow* ($m^3.d^{-1}$).

Q_{peak}: *maximum hourly flow* ($m^3.h^{-1}$). *It is the product of the average flow rate and the peak coefficient (Cp) given by:*

$$Cp_P = 1.5 + \frac{2.5}{\sqrt{Qaverage - L/s)}}$$

The average daily hourly flow is given by: Q_m =/ 24.

The average wastewater flow can be estimated with the following formulas:

$$Q_{average}\,(\frac{m3}{d}) = \frac{Pop.\,L_{person}\,.\,R}{1{,}000}$$

$$Q_{average}\,(\frac{L}{s}) = \frac{Pop.\,L_{person}.\,R}{86{,}400}$$

where:

– Q_m: average flow rate of domestic wastewater ($m^3.d^{-1}$ or $L.s^{-1}$);

– $L_{person\ b}$: water consumption per capita ($L.person^{-1}\ .d^{-1}$);

– R: return coefficient of wastewater flow/water flow.

The water flow rate to be considered is the flow rate actually consumed, not the flow rate produced by the water treatment works. The produced flow rate is higher than that consumed due to unaccounted water losses in the distribution system, which can typically vary from 20 to 50%. Therefore, in an area where the loss is 30%, for every 100 m^3 of water produced, 30 m^3 are not accounted for. Only 70 m^3 are consumed. Of this 70 m^3, about 80% (56 $m^3.d^{-1}$) returns as wastewater into the sewer system.

The average daytime flow corresponds to the 16-hour consecutive daytime period during which the plant receives the greatest volume of wastewater. It generally lasts from 8 a.m. to 12 p.m.

The daytime flow coefficient is the ratio of the average daytime flow to the average daily flow. Statistically, it has been found that 80% of the daily volume is received in 16 hours, the remaining 20% flowing during the off-peak period corresponding to 8 hours.

The average daytime flow is given by:

$$Q_d = Q_d/16$$

In fact, depending on conditions and the volume of discharges, the maximum flow period varies between 2 p.m. and 6 p.m. Depending on the type of urban area, the maximum flow period is determined.

3.2.4. *Maximum and minimum flow rates*

Water consumption and wastewater production in an area vary over the course of the day (hourly variations), the week (daily variations) and the whole year (seasonal variations).

Figure 3.2 shows the typical hourly variations in the influent flow in a WWTP.

Two main peaks can be observed: an early morning peak (more pronounced) and a peak in the early evening (more broadly spread). The mean daily flow corresponds to the line that separates the equal zones, below and above the median line.

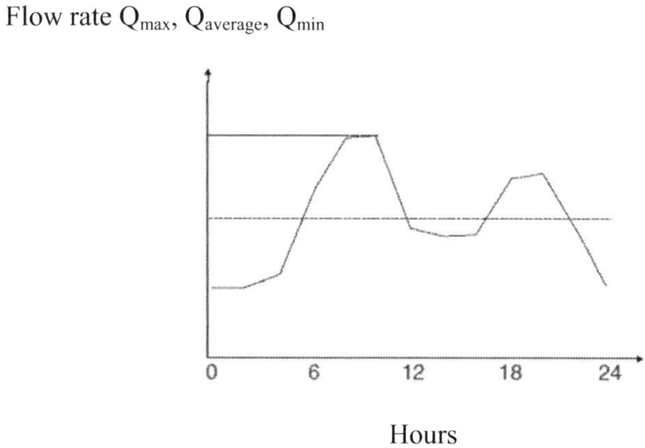

Figure 3.2. *Typical hourly variations in the influent flow in a wastewater treatment plant*

The following coefficients are frequently used to estimate the maximum and minimum flows:

– $K_1 = 1.2$ (peak coefficient where the daily consumption is the highest);

– $K_2 = 1.5$ (peak coefficient where the hourly consumption is most pronounced);

– $K_3 = 0.5$ (hourly reduction coefficient relative to the lowest consumption level).

Therefore, the maximum and minimum flows can be deduced from the following relations:

$$Q_{max} = Q_{average}.K_1\ K_2 = 1.8\ Q_{average}$$

$$Q_{min} = Q_{average}.K_3 = 0.5\ Q_{average}$$

The peak flow of dry weather leads to the definition of a peak coefficient as the ratio of the average flow of the busiest hour to the average daily flow. This peak coefficient can be calculated from the average daily flow Q_m(l/s) by the following formula:

$$Cp = 1.5 + \frac{2.5}{\sqrt{Qm}} \text{ for } Qm > 2.8 \text{ l.s}^{-1}$$

$$Cp = 3 \text{ for } Qm < 2.8 \text{ l.s}^{-1}$$

The peak dry weather flow is given by (Qpdw and Qm are expressed in $l.s^{-1}$):

$$Qpdw = Cp\,Qm = Qm\,(1.5 + \frac{2.5}{\sqrt{Qm}})$$

EXAMPLE.–

– Connected population: 24,000 persons.

– Consumption per capita: $Consump_{person}$ = 160 L/person .d.

– Return coefficient (wastewater flow/water consumption flow): R = 0.8.

The average flow rate is given by:

$$Q_{moy} = Pop.\ Consump_{person}\ .R/1{,}000 = 24{,}000 \times 160 \times 0.8/1{,}000 = 3.072\ m^3.j^{-1}$$
$$(= 35.6\ L.s^{-1})$$

– Maximum flow rate: from Harmon's formula, the ratio $Q_{max}/Q_{average}$ is given by:

$$\frac{Q_{max}}{Q_{average}} = 1 + \frac{14}{4 + \sqrt{P}} = 1 + \frac{14}{4 + \sqrt{(24{,}000/1{,}000)}} = 2.57$$

Therefrom, it is deduced that:

$$Q_{max} = 2.57 \times 35.6\ l/s = 91.5\ L.s^{-1}$$

– Minimum flow rate: by taking a $Q_{min}/Q_{average}$ ratio equal to 0.5, we deduce that:

$$Q_{min} = 0.5 \times 35.6\ L/s = 17.8\ L.s^{-1}$$

3.3. Pollutant loads

3.3.1. *Relationship between load and concentration*

The distribution of pollution forms in typical urban water is as follows:

– $Pollution_{total} \approx 1/3\ Pollution_{soluble} + 1/3\ Pollution_{particulate\ settleable} + 1/3\ Pollution_{particulate\ non\text{-}settleable}$;

– $TSS_{total} \approx 1/3\ SS_{mineral} + 2/3\ TSS_{volatile}$;

– COD/BOD = 2–2.7;

– TSS/BOD = 1–1.5;

– TSS/COD = 0.4–0.7;

– $BOD_{particulate}/VSS = 0.6–1$;

– $COD_{particulate}/VSS = 1.5–2.5$.

The daily flows of pollution in raw urban water are:

– BOD = 55–70 g/PE/day;

– TSS = 60–90 g/PE/day;

– COD = 120–150 g/PE/day;

– NTK = 13–15 g/PE/d;

– P = 2.5–4 g/PE/day.

Raw domestic water can deviate from these values due to industrial water inflow, the length of the network, etc.

In the absence of specific data on concentrations in the different compartments, the method presented in Table 3.2 can be used to estimate the main values.

COD/BOD	< 2			*2–2.7*			> 2.7		
COD_{part}/VSS	1.5			*(1.43 COD/BOD)-1.36*			2.5		
TSS/COD	< 0.4	0.4–0.7	> 0.7	*< 0.4*	*0.4–0.7*	*> 0.7*	< 0.4	0.4–0.7	> 0.7
VSS/TSS	80%	75%	70%	*70%*	*65%*	*60%*	60%	55%	50%

Table 3.2. *Estimation of water compartmentalization (in italics, standard water)*

The load per capita represents the average contribution of each individual (expressed in terms of pollutant mass) per unit of time. A commonly used unit is the gram per inhabitant per day ($g.inhab^{-1}.d^{-1}$). For example, when the contribution of BOD is 54 g.person b.d, this means that each individual discharges an average of 54 grams of BOD per day.

The influx load of a WWTP corresponds to the amount of pollutant (mass) per unit of time. Consequently, the relations are:

Load = population × per capita

Load ($kg.d^{-1}$) = population × per capita load ($g.person^{-1}.d^{-1}$)/1,000 ($g.kg^{-1}$)

Load = concentration × flow rate

Load ($kg.d^{-1}$) = concentration ($g.m^{-3}$) × flow rate ($m^3.d^{-1}$)/1,000 ($g.kg^{-1}$)

The concentration in wastewater can be obtained by the following relation:

Concentration = load/flow rate

Concentration ($g.m^{-3}$) = load ($kg.d^{-1}$) × 1,000 ($g.kg^{-1}$)/flow rate ($m^3.d^{-1}$)

3.3.2. *Pollutant loads in BOD, N and P*

BOD loads provided by raw water per day and per inhabitant are estimated as follows, depending on the type of network:

– separative network: 54 g. $person^{-1}.d^{-1}$;

– pseudo-separative network: 60 $g.person^{-1}.d^{-1}$;

– unified network: 74 g. $person^{-1}.d^{-1}$.

By adopting a discharge of 54 g BOD_5 $person^{-1}.d^{-1}$, the pollutant load becomes:

BOD_5 load = 54 g. person -1.d^{-1} × 24,000 person = 1.296 × 10^6 $g.d^{-1}$ = 1.296 $kg.d^{-1}$

It then follows that:

Average concentration BOD_5 = load/flow rate = (1.296 $kg.d^{-1}$)/(3.072 $m^3.d^{-1}$) = 0.421 $kg.m^{-3}$

Mean concentration BOD_5 = 421 $g.m^{-3}$ = 421 mg. L^{-1}

COD, N and P concentrations can be inferred in the same way.

EXAMPLE.– Calculation of the total nitrogen load in wastewater from the following:

– concentration = 45 mg $N.L^{-1}$;

– flow rate = 50 $L.s^{-1}$.

It follows that:

$$Q = 50\ L.s^{-1} \times 86{,}400\ s.d^{-1}/1{,}000\ L.m^{-3} = 4{,}320\ m^3.d^{-1}$$

The load in N is:

$$Load = 45\ g/m^3 \times 4{,}320\ m^3.d^{-1} 1{,}000\ g/kg = 194\ kg\ N.d^{-1}$$

N loads can be based on values of about 10–18 g/person /day.

In the same order, the total P concentration can be obtained for a load equal to 40 $kg.d^{-1}$:

$$\begin{aligned} Concentration &= 40\ kg.d^{-1} \times 1\ 000\ g.kg^{-1}/4\ 320\ m^3.d^{-1} \\ &= 9.3\ g\ P.\ m^{-3} = 9.3\ mg\ P.L^{-1} \end{aligned}$$

Suspended solid loads are in the 70–90 g/person/day range with about 70% organic matter and 30% mineral matter. This corresponds to a VSS load of 50–64 g/person/d and 20–25 g/person/d for mineral matter.

The following concepts are important in terms of biological wastewater treatment:

– biodegradability: the ability of wastewater to be biochemically stabilized by microorganisms;

– treatability: the ability of wastewater to be treated by physico-chemical or biological processes;

– biodegradable organic matter concentration: wastewater BOD;

– nutrient availability: biological wastewater treatment requires a balance between C, N and P nutrients;

– toxicity: some urban wastewater may be contaminated with toxic organic substances or with toxic or inhibitory elements (Ag, As, Cd, Co, Cr, Cu, Hg, Ni, Pb, Sb, Se and Zn), which may affect biological treatment.

3.4. References

Brix, H. (ed.) (1993). Wastewater treatment in constructed wetlands: System design, removal processes, and treatment performance. In *Constructed Wetlands for Water Quality Improvement*. CRC Press, Boca Raton, 9–22.

van Buuren, J.C.L, Frijns, J.A.G., Lettinga, G. (1995). *Wastewater Treatment and Reuse in Developing Countries*. Wageningen Agricultural University, Wageningen.

Campos, H.M. and Von Sperling, M. (1996). Estimation of domestic wastewater characteristics in a developing country based on socio-economical variables. *Water Science and Technology*, 34(3/4), 71–77.

Chagnon, F. and Harleman, D.R. (2002). *An Introduction to Chemically Enhanced Primary Treatment*. Massachusetts Institute of Technology, Cambridge.

Environmental Protection Agency (EPA) (1985). *Rates, Constants, and Kinetics Formulations in Surface Water Quality Modeling*. EPA, Athens.

Environmental Protection Agency (EPA) (1987). *Design Manual. Dewatering Municipal Wastewater Sludges*. EPA, Cincinnati.

Environmental Protection Agency (EPA) (1998). *Wastewater Treatment Works... The Basics*. EPA, New York.

Gandhi, V. and Shah, K. (2021). *Advances in Wastewater Treatment I*. Materials Research Foundations, Millersville.

Gedda, G., Balakrishna, K., Devi, R.U., Shah, K.J. (2021). Introduction to conventional waste-water treatment technologies: Limitations and recent advances. In *Advances in Wastewater Treatment I*, Gandhi, V., Shah, K. (eds.). Materials Research Foundations, Millersville, 1–36. doi: 10.21741/9781644901144-1.

Guyer, J.P. (2018). *An Introduction to Preliminary Wastewater Treatment*. Guyer, New York.

Lekang, O.I. and Kleppe, H. (2000). Efficiency of nitrification in trickling filters using different filter media. *Aquacultural Engineering*, 21(3), 181–199.

Marais, G.R. and Ekama, G.A. (1976). The activated sludge process. Part 1 – Steady state behaviour. *Water S.A.*, 2(4), 164–200.

Metcalf & Eddy, Inc. (2004). *Wastewater Engineering, Treatment and Reuse*. McGraw-Hill, New York.

Nathanson, J.A. and Ambulkar, A. (2025). Wastewater treatment. *Encyclopaedia Britannica*.

Pennsylvania Department of Environmental Protection (2016). Module 1: Introduction to waste-water treatment. Revised 2016. Pennsylvania Department of Environmental Protection, The Pennsylvania State Association of Township Supervisors (PSATS), Gannett Fleming, Inc., Dering Consulting Group, Penn State Harrisburg Environmental Training Center.

Ramalho, R.S. (1977). *Introduction to Wastewater Treatment Processes*. Academic Press, New York.

Tatoulis, T.I., Tekerlekopoulou, A.G., Akratos, C.S., Pavlou, S., Vayenas, D.V. (2015). Aerobic biological treatment of second cheese whey in suspended and attached growth reactors. *Journal of Chemical Technology & Biotechnology*, 90(11), 2040–2049.

Von Sperling, M. (2007). *Basic Principles of Wastewater Treatment: 2*. International Water Association, New York.

Von Sperling, M. and Chernicharo, C.A.L. (2002). Urban wastewater treatment technologies and the implementation of discharge standards in developing countries. *Urban Water*, 4, 105–114.

Von Sperling, M. and Goncalves, R.F. (2002). Sludge production in wastewater treatment systems in warm-climate countries. In *3rd IWA World Water Congress*. International Water Association, Melbourne, 7–12 April.

4

Pretreatments

Physical pretreatment consists of a series of operations for reducing the coarsest materials in raw water or those that might affect the subsequent treatment process.

The physical pretreatment operations are:

– screening;

– comminution;

– sand removal;

– grease and oil removal.

4.1. Screening

Materials such as rags, plastic and alien materials can significantly interfere with processing processes or damage plant equipment if they are not removed. These materials are removed by using a sieve. The examined materials are hazardous and must be treated and disposed of to avoid any human health problem, mechanical failures and to limit the production of unpleasant odors. Screens can be mechanical or manual. They operate by separating the largest materials from the raw water by directing the influent through bars in which the spacing between them is determined according to the nature of the effluent:

– Manual screens: low to no equipment maintenance and a good alternative for small plants with little projection, but should be cleaned regularly to prevent debris build-up.

– Mechanical screens: reduced labor costs and improved flow conditions and sieving capture, but may require high maintenance for the cooling equipment. Require a constant supply of energy.

The performance of a screen (manual or self-cleaning) is characterized by its spacing between the bars. Spacing between the screen bars is defined by the choice of the size and nature of the objects accepted by the plant. A trade-off is also looked for between bar spacing and the quantity of waste to be evacuated (frequent cleaning of the screen).

A distinction is made between:

– pre-screening: 30–100 mm spacing;

– medium screening: 10–25 mm spacing;

– fine screening: 3–10 mm spacing.

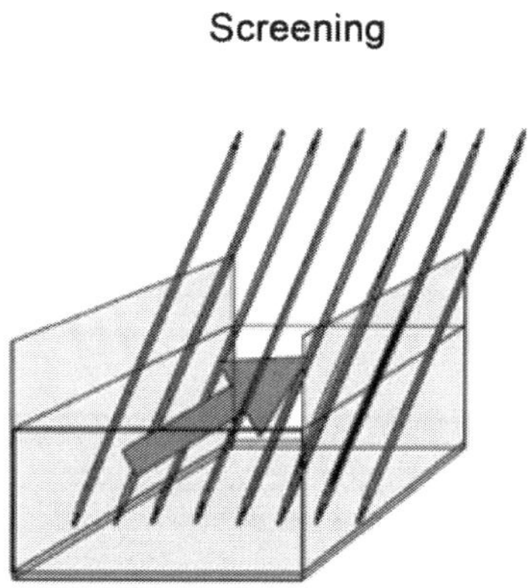

Figure 4.1. *Schematic diagram of a bar screen*

There are different types of screens depending on the design of the manufacturers and the nature of the effluent to be treated.

4.1.1. *Manual screens*

Consisting of bars most often inclined at 60–80° with respect to the horizontal, they are, however, preferred for very small plants. Cleaning is done using a rake. The disadvantage of these screens is that they need to be cleaned daily because if the screen remains uncleaned for a long time, the effluent may flow back up into the inlet collector.

4.1.2. *Mechanical screens*

With more than 2,000 population equivalents, the plant must be equipped with mechanical screens. This mechanization is essential to prevent quick clogging of the pipes.

Mechanical screens are classified into two categories:

– straight screens;

– curved screens.

4.1.2.1. *Straight screens*

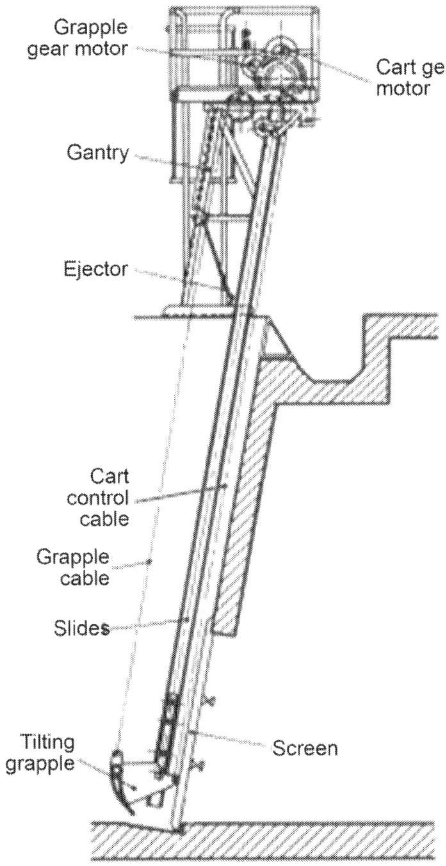

Figure 4.2. *Straight mechanical screen with a scraper*

The raw water passes through screens consisting of bars placed vertically or inclined at 60–80° with the horizontal. The space between bars varies from 6 to 100 mm. The average speed of flow between bars is between 0.6 and 1 $m.s^{-1}$.

The screen is usually cleaned mechanically. This is achieved by a rake attached to a carriage which moves from bottom to top along a rack or driven by two cables (see Figure 4.1). It can be controlled by a timer and/or from a differential pressure drop indicator.

When they are strongly tilted (80° inclination), they are designed with different cleaning devices such as:

– rakes or combs;

– brushes installed on an endless chain;

– alternative grapples, controlled by cables, making it possible to convey the litter up to great heights.

They can be installed upstream or downstream with a 4–6 cm bar spacing (see Figure 4.2).

The automatic operation of the system is controlled by an electric clock or even a pressure drop indicator. Screening materials are an unwanted product that is composted or disposed of by truck.

4.1.2.2. *Curved screens*

Figure 4.3. *Automatic inclined screening*

These screen racks are designed to treat the water of a plant treating 10–5,000 $m^3.h^{-1}$. Consisting of flat iron bars disposed in a quarter circle, they are cleaned by a double rotating rake or by a system of connecting rods applied against the screen (see Figure 4.4).

Figure 4.4. *Curved screen*

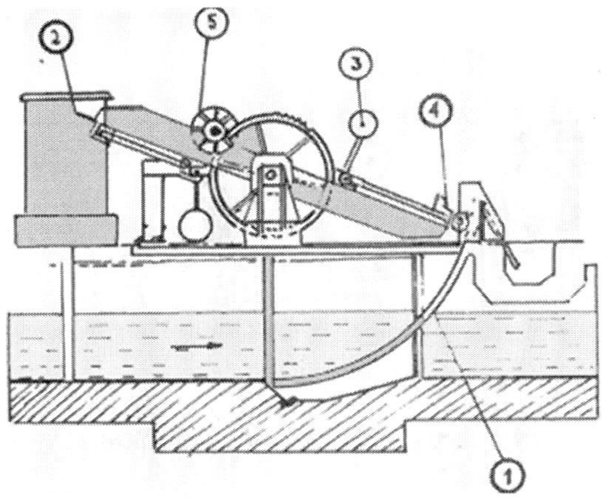

Figure 4.5. *Curved mechanical screen: 1) curved screen; 2) comb; 3) counterweight; 4) drainage; and 5) drive motor*

4.1.3. *Screen calculation*

Kirschmer (1926) established a relationship between the load loss, the shape coefficient of the bars and the angle of the screen with respect to the horizontal:

$$\Delta h = \beta \left(\frac{l}{b}\right)^{4/3} hv \sin\alpha$$

where Δh is the load head loss; β is the bar shape coefficient; 1 is the apparent width of the bars; b is the spacing between bars; hv is a term in $v^2/2g$ upstream of the bars; and finally α is the angle of the bar rack with the horizontal.

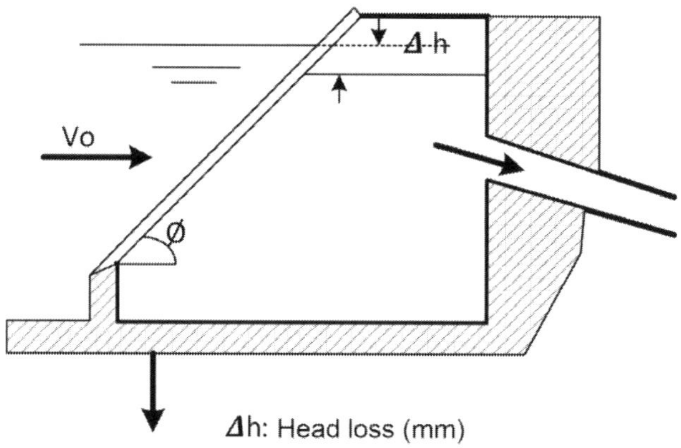

Figure 4.6. *Screen load loss coefficient*

The values of the thickness of the bars (β) depend on the shape of the bars (Table 4.1).

Bar types	B
Rectangular cross-section	2.42
Rectangular cross-section with semi-circular rounding upstream	1.83
Rectangular cross-section with semi-circular rounding upstream and downstream	1.67
Circular cross-section	1.79
Ovoid cross-section with a large width upstream	0.76

Table 4.1. *β values according to bar type*

The typical values for the selection of a manual or automatic screen are presented in Table 4.2.

Parameters	Unit	Cleaning mode	
		Manual	Automatic
Bar size			
Width or thickness	mm	5–15	5–15
Depth	mm	25–38	25–38
Air gap	mm	25–50	15–75
Tilt angle	°	30–45	0–30
Maximal			
Velocity	$m.s^{-1}$	0.3–0.6	0.6–1.0
Minimal	$m.s^{-1}$	–	0.3–0.5
Acceptable load losses	mm	150	150–600

Table 4.2. *Typical values for selecting screens*

Figure 4.7 allows us to calculate the load loss in the screen using Bernoulli's equation.

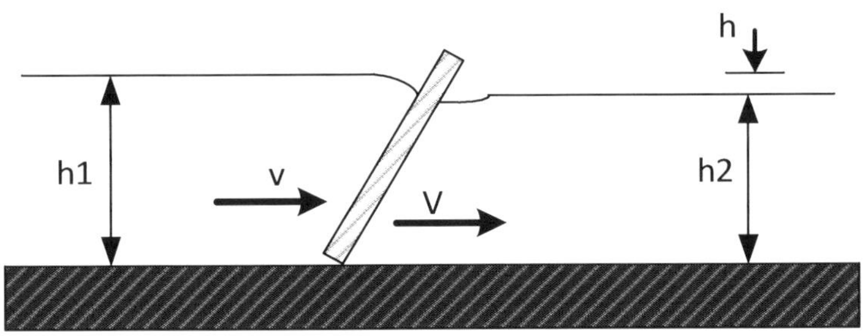

Figure 4.7. *Wastewater flow through the screen*

We have:

$$h1 + \frac{v^2}{2g} = h2 + \frac{V^2}{2g} + \Delta h$$

such that:

$$h = h1 - h2 = \frac{V^2 - v^2}{2gC^2}$$

where:

– h1: downstream water depth (m);

– h2: upstream water depth (m);

– h: load head loss (m);

– V: velocity through the screen ($m.s^{-1}$);

– V: estimated velocity in the downstream screen channel ($m.s^{-1}$);

– g: gravitational constant, 9.81 m/s^2;

– C: discharge coefficient with a typical value of: $C^2 = 0.7$ for a clean screen; $C^2 = 0.6$ for a clogged screen.

Other formulas are offered by the manufacturers. For very high flow rates, the following calculation is indicated, for a straight fine screen:

$$L = \frac{d + e}{e} \frac{1}{1 - \eta} \frac{Qpdt}{v\,h}$$

where:

– L: total screen width (m);

– d: bar width (mm);

– e: spacing between bars (mm);

– η: soiling degree;

– Qpdt: peak dry time flow ($m^3.s^{-1}$);

– h: water level upstream of the bar racks (m);

– V: flow velocity through the racks ($m.s^{-1}$).

$$S\ screen = \frac{hmax\,l\,(1 - \beta)\delta}{\sin\alpha}$$

where:

– β: fraction of the area occupied by the bars;

– δ: screen clogging coefficient; it is equal to 0.5 for automatic screening and 0.25 for manual screening;

– hmax: maximum permissible water height on a screen of width 1;

– α: bar rack angle of inclination.

Other formulas are used by manufacturers such as:

$$W = \frac{Q}{vD}\frac{B+S}{S} + C$$

where:

– W: screen width (m);

– Q: maximum flow rate ($m^3.s^{-1}$);

– v: speed through the bars ($m.s^{-1}$);

– D: water flow height (m);

– B: bar width (m);

– S: spacing between bars (mm);

– C: coefficient linked to the general framework of the rack.

EXAMPLE.– For a 20 mm spacing between bars and a 10 mm bar thickness, the width of the bar rack to be adopted for a flow rate of 97.1 $m^3.h^{-1}$ and a flow velocity of 1 $m.s^{-1}$ is obtained as follows:

– maximum water height = 0.15 m;

– flow area: S = Qpdt/V;

– surface area = 97.1/(3.600 x 1) = 0.0269 m^2;

– β is equal to 1/3, that is (1 – β) = 2/3 = 0.66;

– δ = 0.50.

The width is deduced as:

$$l = \frac{0.0269 \text{ x } \sqrt{3}}{2 \text{ x } 0.66 \text{x } 0.50 \text{ x } 0.15} = 0.47 \text{ m}$$

EXAMPLE.– Computation of the air gap and screen dimensions:

Sesp = total spacing area through the screen;

Qpdt = peak dry time flow ($m^3.s^{-1}$)/velocity ($m.s^{-1}$);

– Peak flow: 500 m^3/h = 0.139 $m^3.s^{-1}$.

– Total width of the air gap through the screen:

$l_{airgap} = Sesp/h1;$

$l_{airgap} = 0.231\ (m^2)/0.6\ (m);$

$l_{airgap} = 0.385$ m.

– Spacing between bars: b = 50 mm.

– Number of spacings:

$n = l_{airgap}/b;$

$n = 0.385/0.05;$

$n = 7.7 = 8$ bars;

$l_{corrected\ airgap}$: $8 \times 0.05 = 0.40$ m.

The bars are chosen such that the width is 15 mm and the depth is 38 mm.

– Channel width:

$l_{channel} = 0.40 + (0.015 \times 8);$

$l_{channel} = 0.52$ m.

– Screen height:

$H = h1/\sin 45^o;$

$H = 0.6/0.707;$

$H = 0.85$ m.

That is, a pipe depth equal to $5 \times 0.85 = 4.25$ m.

– Velocity between bars through the screen:

$Sb \times v = Aa \times V;$

$V = Sb\ v/S.$

– Area between bars:

$(0.015 \times 8)1.5 = 0.18\ m^2;$

$v_{bars} = Qpdt/S_{bars} = 0.139/0.18 = 0.77\ m.s^{-1}$.

– Screen efficiency:

$Eff. = (l_{airgap}/l_{channel}) \times 100;$

$Eff. = (0.385/0.52) \times 100 = 74\%.$

– Load loss through the screen:

$$h = h1 - h2 = \frac{v\ bars^2 - v^2}{2gC^2} = \frac{0.77^2 - 0.6^2}{2 \times 9.81 \times 0.7^2} = 0.0242\ m$$

4.1.4. *Screen implementation*

Between bars, velocities between 0.50 and 1.2 m/s are generally adopted. The spacing between bars will be chosen according to the nature of the water to be treated and the quality of the desired screening. Bar thickness must be as small as possible, and the ratio of bar gap/spacing + thickness should be as large as possible.

The operation of the plant will be less demanding if it is possible to alternate between two mechanical screens operating in parallel. The operation time of a screen is about three to four hours per day except in the peak season during which they can operate for 24 hours.

A channel for bypassing the screen is usually provided, equipped with a manual relief screen for repair maintenance of the main screen.

As a first approximation, the volume of the collected waste is given by:

V (l/person.year) = 12–15/e

where e is the spacing between bars (cm).

4.2. Comminution

This is an operation for crushing the waste that is evacuated with the raw water. This system has many requirements depending on the nature of the waste and the effectiveness of this treatment. There are two types of crushers: hammer crushers and rake drums.

Using comminution is expensive for a plant because of the difficult maintenance of this system, and also because of the consequences that can derive from a malfunction (pipe clogging, pump clogging).

4.3. Sieving

Sieving is in fact an advanced screening process and consists of cloth filtration using meshes of different sizes. There is macrosieving (meshes greater than 0.3 mm) and microsieving (meshes less than 100 μ).

Macrosieving is the most important step and is intended to block floating suspended solids, plant litter, grasses, insects, etc. This reduces the pollution load and alleviates the wastewater treatment plant.

There are two types of macrosieving devices:

– pressure washing sieves;

– sieves with mechanical waste evacuation.

4.3.1. *Pressure washing sieves*

The device comprises a horizontal axis drum with a diameter of several meters, or it can include a rotating belt over an endless chain. These are therefore rotating macrosieves.

Sieving is carried out on filter panels equipped with a 0.3–3 mm metal mesh. The washing water conveys the retained waste towards an internal collection and discharge chute.

Mechanical sieving filters can also be found that operate under pressure fed by pumping. The mesh range spans between 0.15 mm and 4 mm.

4.3.2. *Mechanical waste discharge sieves*

There are cleaning sieves that operate by scraping that allow sieving water after conveying it through a corrugated and perforated sheet, with 2–5 mm holes. The length of the sheet is about 2 m.

Another technique consists of circular or rectangular vibrating screens whose printed shaking ensures the removal of waste.

4.4. Sand removal

Pretreatment is intended for the disposal of:

– coarse solids;

– particles > 1 cm;

– sand.

The purpose of sand removal is to remove heavy inorganic matter, which easily settles, from the wastewater in order to prevent damage to the mechanical installation or sedimentation in downstream installations. Grit can include sand, silt, glass, small stones and other large organic and inorganic matter (litter). Excess grit can cause operational problems such as pump blockages and high levels of organic matter in digesters and/or reactors.

The basic removal mechanisms are physical.

Figure 4.8 shows a typical treatment regimen for preliminary treatment.

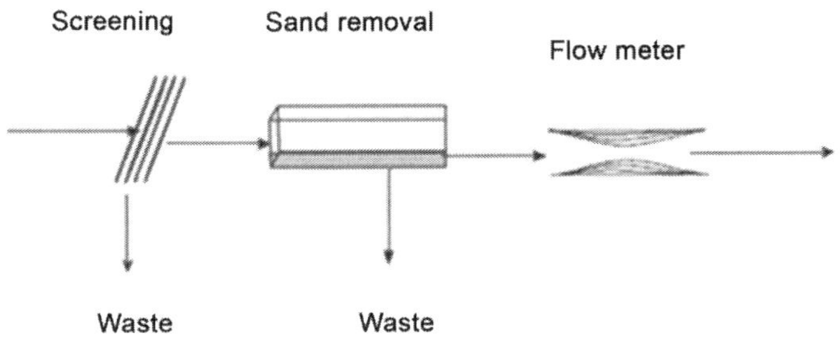

Figure 4.8. *Typical pretreatment diagram*

Sand removal is carried out on particles larger than 200 μm. The sedimentation rate is calculated by Stokes's law (free fall). We calculate the section of the grit trap such that the water velocity does not fall below 0.30–0.20 m/s; this thus prevents organic matter from being deposited at the same time as the sand.

The structures to be planned for obtaining a flow velocity of 0.30 m/s are, in order of importance:

– channel grit traps (with straight flow), whose flow velocity is variable or constant;

– circular grit traps with tangential feed or mechanically stirred or an air blower (to avoid the sedimentation of organic matter, during off-peak hours, with low flow rate);

– rectangular air-blower grit traps. Air is blown in which causes the liquid to rotate and thus creates a constant sweeping speed of the bottom, perpendicular to the transit velocity, which, being much lower, can then be variable without inconvenience. The sand is extracted either mechanically by scraping at a reception plant then pumped back or directly by a suction pump mounted on an overhead crane.

The separated sand contains organic matter, and several devices are applied to improve its quality, such as hydrocyclonic washing or extraction from the storage pits by mechanical means, which at the same time operate as sand washers (angled troughing Archimedes' screw, reciprocating motion classification, etc.). The volume of sand extracted per person per year is about 5–12 dm^3.

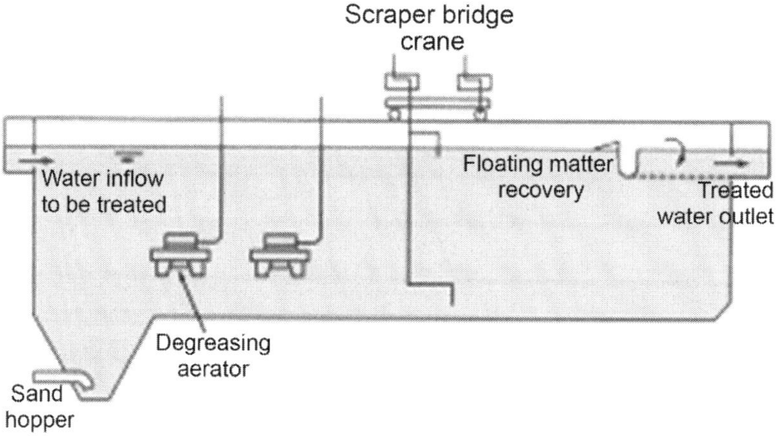

Figure 4.9. *Combined sand and oil removal*

The flow through the sand separator is oriented in such a way in order for the sand to settle and to keep light organic matter in suspension. For aerated grit removal, this effect is created by aeration with large air bubbles. Sometimes, sand and oil removal are combined. This will thus ensure that non-emulsified fat will float to the surface. The sand is continuously removed by way of scrapers and air lift pumps or periodically by manually shoveling the sand. The grease is removed by a surface scraper. A sand washing plant scrapes the sand and deposits it in a container. Washing the sand reduces any unpleasant odors emitted by the organic material.

Comminution is the reduction of solids from a medium particle size to a smaller particle size by crushing, grinding, cutting, vibration or other processes.

The removal of coarse solids is frequently performed using sieves or bar racks, but static particles or rotary screens and crushers can also be used.

There are coarse, medium and fine screens, depending on the spacing between bars. The retained materials can be manually or mechanically removed.

The main goals of coarse solid removal are:

– protect wastewater transport equipment (pumps and pipes);

– protect equipment in downstream treatment structures;

– avoid sand accumulation in areas with little mixing;

– prevent corrosion in air diffusers in aeration lagoons.

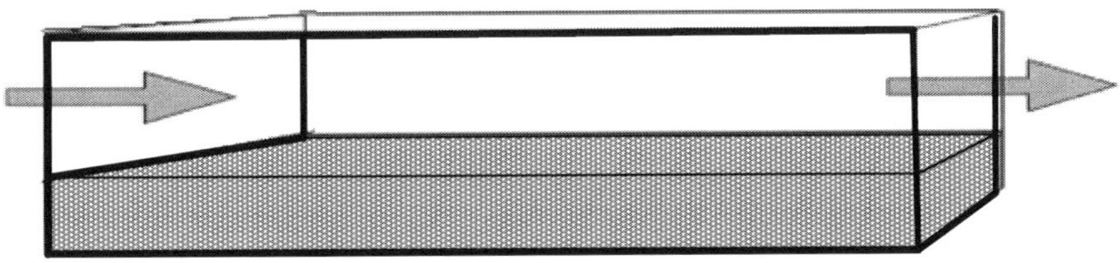

Figure 4.10. *Sand trap schematic diagram*

The sand removal mechanism is simple in sedimentation: the sand grains fall to the bottom of the tank due to their greater density, while the particles remain suspended and are directed to the downstream units.

The flow velocity that allows the separation of organic matter and sand is 0.3 $m.s^{-1}$. Typically, the required air flow is 4.6–7.7 $l.s^{-1}$ per length meter to facilitate de-oiling. The transverse velocity at the water surface must be between 0.6 and 0.8 $m.s^{-1}$. The hydraulic retention time on the peak flow should be between 3 and 5 min.

Removing sand from the raw effluent is essential when we want to protect pipes and pumps from corrosion and also to prevent clogging in pipes by sedimentation during treatment.

Figure 4.11. *Screening and sand removal*

The rate of particle sedimentation depends on their nature, their diameter and the viscosity of the liquid in which they are found. Table 4.3 gives an idea of the sedimentation rate of some materials.

Materials	Diameter (mm)	Specific weight ($g.cm^{-3}$)	Sedimentation rate ($cm.s^{-1}$)
Fine sand	0.02–2.0	2.60	2.00-100
Silt	0.002–0.02	2.60	0.04–2.0
Clay	0.002	2.60	0.005–0.04
Organic matter	0.01–0.4	1.01	0.03–0.2

Table 4.3. *Physical characteristics of various materials*

There are various types of sand traps, including:

– channel grit traps;

– square grit traps;

– aerated grit chambers.

These can be single grit channel traps, which are channels with wide, rectangular cross-sections, in which the flow velocity is less than 0.3 $m.s^{-1}$ in order to avoid particles being carried away by the flow of water.

There are also constant-velocity channel grit traps whose disadvantage is that they are sensitive to daily variations in flow rate and therefore reintroduce the deposited particles back in motion. Square sand traps with flat bottoms have been imagined, where water covers the entire width of the tank. The water leaves by the opposite side, and the sand is collected by scraping.

Figure 4.12. *Longitudinal grit trap (© Estruagua, Spain)*

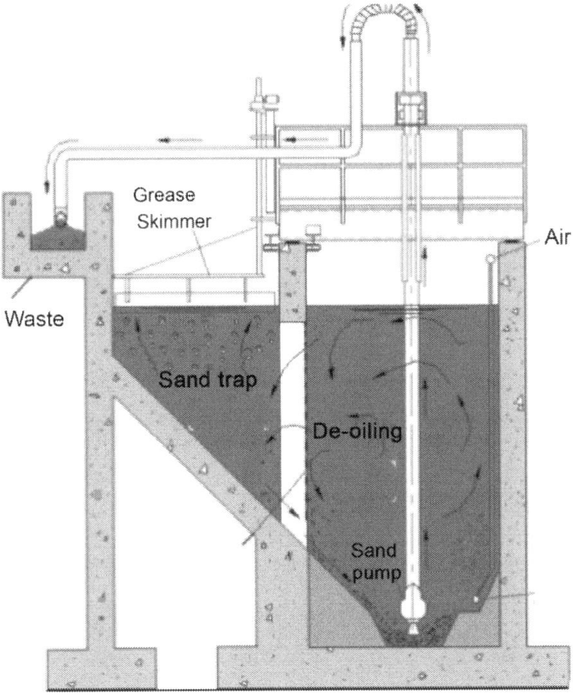

Figure 4.13. *Aerated sand and grease trap (© AES USA)*

In aerated sand traps, a rotating motion is induced in water by blowing air, which:

– settles sand grains;

– cools down the water by feeding oxygen.

The retention times are 3–5 min, and the air injection is 1–1.5 $m^3.m^3$ of water. There are also circular grit traps that are tangentially fed and in which sand is projected towards the wall and goes downward in a vertical spiral. The velocity flow is about 0.8 $m.s^{-1}$, and the retention time is about 1 min.

The volume of sand collected in the grit traps varies between 5 and 12 liters per person per year. It depends on the nature of the effluent, the type of network, the state of the road, etc.

In this type of structure, the flow entering the weir is given by:

$$Q = K\ l\ H^{3/2}$$

where H is the load at the weir inlet, l is the weir width and K is a coefficient depending on the characteristics of the weir. We take K = 0.4 for a thin-walled rectangular weir. In addition, in the sand removal channel, the flow rate is:

$$Q = S.V$$

where S is the cross-section of the sand trap and V is the flow velocity of water.

The point is to keep a constant V, which is the same as keeping Q/S constant. We have:

$$K\ l\ H^{3/2} = a$$

For an arbitrary cross-section channel $S = \int_0^H l'dH$ where l′ is the width of the channel, we have:

$$S = a \int_0^H l'dH = K\ l\ H^{3/2}$$

$$a\ l'dH = 3/2\ K\ l\ H^{1/2}\ dH$$

$$H = (\frac{2\ l'a}{3\ K\ l})^2$$

which requires a parabolic channel cross-section.

When the channel cross-section is rectangular, the flow rate is directly proportional to the load height with Q = H l′ V.

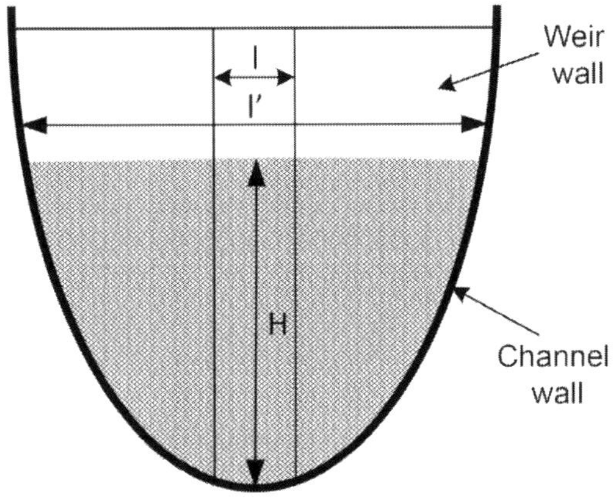

Figure 4.14. *Schematic cross-section of a horizontal-flow sand trap*

The calculation of the length of the horizontal-flow sand trap channel is obtained by:

$$H\ max/v_c = t = L/v$$

where H max is the height in the channel for Q maximum, v_c is the sedimentation velocity of the smallest particles (about 2.3 $cm.s^{-1}$), t is the retention time (s), v is the flow velocity (30 $cm.s^{-1}$) and L is the length of the tank.

EXAMPLE.– Data (Maystre 1973):

– horizontal velocity to be controlled: 30 $cm.s^{-1}$ of rectangular weir;

– flow rates:

- 5,000 $m^3\ d^{-1}$ (maximum rate),
- 2,500 $m^3.d^{-1}$ (average rate),
- 1,000 $m^3.d^{-1}$ (minimum rate);

– channel width: 2 m (see Figure 4.14);

– water height (H) for the maximum flow rate:

$$S = 2/3\ H\ l' = Q/v$$

Let:

$H = 3/2\ Q/v\ l'$

$$H = \frac{3}{2}\,\frac{5{,}000\ l}{86{,}400 \times 2 \times 0.3} = 0.144 \text{ m}$$

– weir cross-section:

$$H + \frac{v^2}{2g} = hc + \frac{vc^2}{2g} + 0.1\frac{vc^2}{2g}$$

Since:

$hc = 2\ vc^2/2g$

where vc is the critical velocity.

Let:

$$H + \frac{v^2}{2g} = 2\frac{vc^2}{2g} + \frac{vc^2}{2g} + 0.1\frac{vc^2}{2g} = 3.1\ \frac{vc^2}{2g}$$

Assuming that $v^2/2g$ is negligible and = 0.004, it follows that:

$$hc = 2\frac{vc^2}{2g} = 2\frac{H}{3.1} = 0.092 \text{ m}$$

$vc = (g.\ hc)^{1/2} = (0.092 \times 9.81)^{1/2} = 0.95\ m.s^{-1}$

Let:

$Q/vc = 500/(0.95 \times 86{,}400) = 0.061\ m^2 = S$

$L = S/hc = 0.061/0.092 = 0.66$ m

– Calculation of the channel width and the draught for the other flow rates:

- for the average flow rate:

$$S = \sqrt[3]{\frac{2{,}500^2\ 0.66}{86{,}400^2\ 9.81}} = 0.038 \text{ m}^2$$

hc = 0.038/0.66 = 0.58m

H = (3.1 x 0.058)/2 = 0.086 m

l′ = 3 Q/2 H v = (3 x 2 500)/(2 x 86,400 x 0.086 x 0.3) = 1.68 m

- for the minimum flow rate:

$$S = \sqrt[3]{\frac{1{,}000^2\ 0.66}{86{,}400^2\ 9.81}} = 0.02\ \text{m}^2$$

hc = 0.02/0.66 = 0.031 m

H = (0.031 x 3.1)/2 = 0.05 m

l′ = (3 x 1,000)/(2 x 86,400 x 0.05 x 0.30) = 1.15 m

– Calculation of the lagoon length: for a maximum flow rate, the contact time is:

t = H/vc = 0.144/0.023 = 6.23 s

L = v.t = 0.3 x 6.23 = 1.86 m

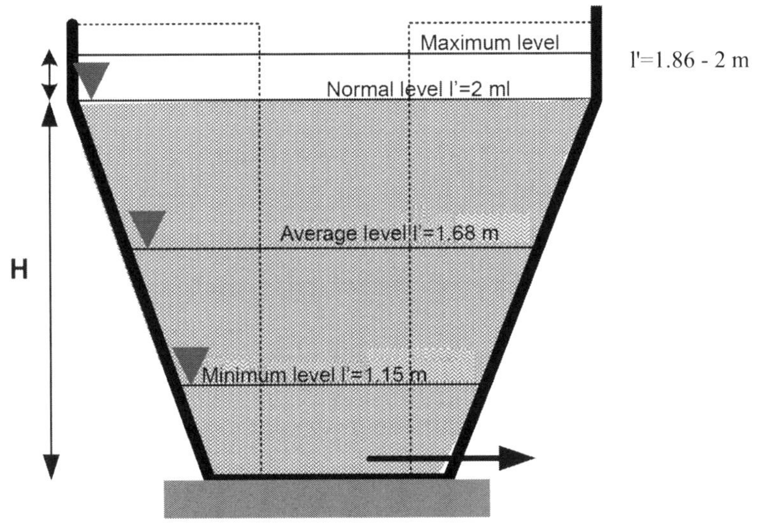

Figure 4.15. *Sand trap channel cross-section*

4.4.1. *Calculation for aerated sand trap*

The maximal flow rate to be treated by the treatment plant is 397 $m^3.h^{-1}$. We admit the following assumptions for the calculation: a hydraulic retention time for the peak flow of five minutes. Both operations are simultaneously carried out in a combined rectangular structure.

– *Step 1: tank volume.*

$397 \times (5/60) = 33.1\ m^3$

– *Step 2: tank dimensions.*

Let us suppose a 2 m width. Consider the depth-to-width ratio of 1.5:1 (typically from 1.5:1 to 2:1):

- depth: $2 \times 1.5 = 3$ m;
- length: $33.1/(2 \times 3) = 5.5$ m.

Therefore, the tank is 5.5 m x 2 m x 3 m = 33 m^3.

Let us calculate the amount of air needed. It is assumed that the required airflow is 4.6 l/s/m. The total airflow is:

$4.6 \times 5.5 = 25.3\ l.s^{-1}$

– *Step 3: amount of sludge recovered.*

The quantity of sludge is considered to be 0.015 $l.pers^{-1}.d^{-1}$ and the quantity of grease 0.1 kg per 1,000 persons per day. Then:

- sludge volume: $0.015 \times 23{,}000 \times 24 \times 0.001 = 8.28\ m^3$;
- grease mass: $0.1 \times (23{,}000/1{,}000) \times 24 = 55.2$ kg.

4.5. Oil and grease removal

This operation is intended to reduce non-emulsified greases and oils by simple physical surface sedimentation.

Oils and greases have many disadvantages for the subsequent biological treatment, such as poor oxygen diffusion in the bacterial floc, pump and pipe clogging and acidification of the environment in the anaerobic digester with all the consequences that this can represent.

Oils and greases are mainly derived from their use or preparation in the food sector. The presence of oils and greases has several consequences for the pumping plant and wastewater treatment:

– grease accumulation and thickening on the walls of the wet well;

– detaching of large pieces of grease from the walls and disruption of operations due to obstruction in the piping on the suction side of the pump or the pump itself;

– foam formation on the surface of the tanks;

– bar rack obstruction;

– deposit formation on the walls of the tanks;

– degradation of the paint on surfaces with which oils and greases are in contact;

– appearance of unpleasant odors when decomposing oils and greases.

Old plants addressed grease removal solely using a static phenomenon by exploiting the difference in density between greases, oils and the medium.

There are different de-oiling–grease removal devices designed according to the nature of the water to be treated and the concentrations to be removed.

4.5.1. *Grease separators*

Also known as “grease boxes”, these are designed for flow rates of 20–30 $l.s^{-1}$. Short retention times (three to five minutes) and an ascent rate of 15 $m.h^{-1}$ have led garages and filling stations to adopt them. The efficiencies obtained are 80–90%.

4.5.2. *Aerated grease and oil trap*

These operations concern a large volume of wastewater created by food industries (delicatessens, meat canning factories, slaughterhouses, margarine plants, oil mills, etc.), as well as from oil refineries and mechanical industries.

It can be considered that oil and grease removal involves the extraction of any floating matter with a density lower than that of water. These materials are of a very diverse nature, and their quantities are estimated by measuring “solvent extractable matter”. The content of some industrial discharges (slaughterhouses, dairies, etc.)

can reach values of 300–500 mg. L^{-1}; moreover, these materials may form a stable emulsion maintained by mixing water or constitute an independent non-emulsified phase.

Oils and greases, when not emulsified, are separated as a floating sludge in longitudinal or circular structures with an aeration zone where air bubbles increase the rising rate of fat particles and a stabilization zone where recovery takes place (see Figure 4.16).

This type of structure is calculated for a maximum upflow velocity of 15 $m^3.m^{-2}.h^{-1}$ and retention times of 8–15 min. The blown air flow rate is approximately 0.3 m^3 per cubic meter of water per hour.

In the petrochemical industry, in the case of waters loaded with hydrocarbons, American Petroleum Institute (API) separators are successfully used, which consist of rectangular longitudinal circulation tanks, in which the bottom and the surface are scraped by scrapers mounted on endless chains. Several compartments can be associated in series.

If the emulsion is not too fine (particles greater than 50 μm), the removal of emulsified oily materials can be ensured by gravity separation using Parallel Plate Interceptor (PPI) oil separators in which the oil droplets only have to travel a short path before being captured, limited by the spacing between plates (variable from 20 to 100 mm).

These devices exhibit high compactness and efficiency with fine droplets of wastewater with a medium hydrocarbon content. It should be noted that, for stable oily emulsions, hydrocarbon separation cannot be achieved unless the emulsion is first “broken” by chemical means (by feeding strong acids, coagulating reagents such as iron or aluminum salts).

This type of structure includes an aerated zone (with bottom insufflation) followed by a lateral sedimentation compartment calculated for an ascent rate of 15–20 $m.h^{-1}$.

Retention times are 3–10 minutes computed from the maximal flow rate. The products are recovered on the surface by means of manual or mechanical scrapers. The air flow to be introduced depends on the concentration of greases and oils, and is about 5–10 $m^3.h^{-1}$ and per m^3 of the structure capacity. The pressure of the injected air is approximately 0.5 bar.

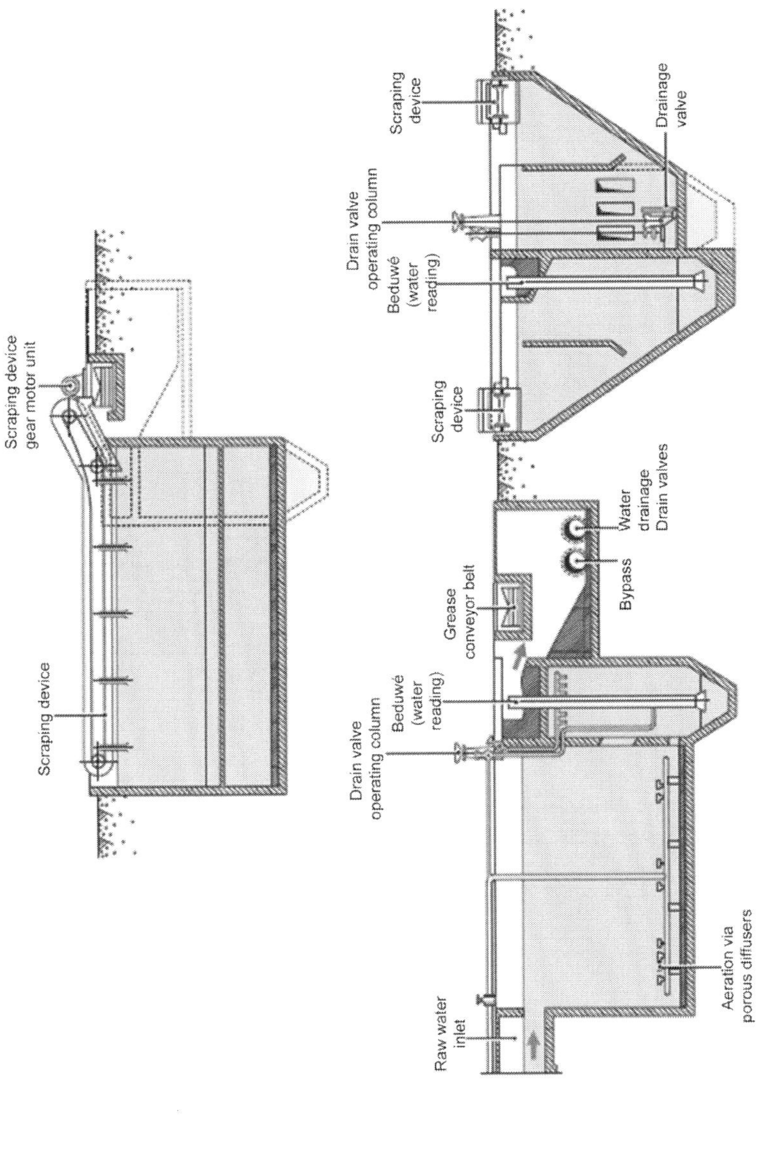

Figure 4.16. *Aerated oil and grease removal longitudinal unit (Degrémont 2004)*

4.5.2.1. *Longitudinal oil remover*

This is a rectangular tank equipped with distributors and scrapers on the surface and at the bottom. The structure is calculated for an ascent velocity of 0–3.6 $m.h^{-1}$, for horizontal flow velocity of 20–50 $m.h^{-1}$, a 2–6 m width and a 1–3 m water height. This is a natural flotation system allowing excellent performance and low maintenance.

It is atmospheric air, dissolved air or vacuum flotation that is the main focus of manufacturers when oil and grease concentrations are high (connection of effluents from slaughterhouses, rendering plants, refineries, etc.).

4.5.2.2. *Calculations for air flotation facilities*

Without recycling, the relationship between air pressure and the air/solids ratio is:

$$\frac{a}{X} = \frac{s\,K(fP - 1)}{X'}$$

where:

– a/X is the air/product ratio ($cm^3.mg^{-1}$);

– K is a coefficient depending on the surface properties of the products to be eliminated (0.8–1.3);

– f is a factor representing the fraction of dissolved air at pressure P, which is therefore dependent on the quantity of air to be applied per unit volume of water and the operational time;

– P is the pressure (atm);

– X, is the concentration of products to be eliminated ($mg.L^{-1}$);

– s is the solubility of air in water under 1 atm ($cm^3.L^{-1}$).

If recycling is included, we obtain:

$$\frac{a}{X} = \frac{s\,K(fP - 1)\,R}{QX'}$$

where R is the recycling flow rate ($m^3.s^{-1}$) and Q is the effluent inlet flow rate ($m^3.s^{-1}$).

The values of f often range between 0.2 and 0.9, while those of a/X fall in the 0.005–0.060 range. For a chosen value of a/X, pressure P can be calculated and then R derived, if necessary.

The surface area of the structure is given by the relation S = Q/Vf where Vf is the flotation velocity.

As an example, Table 4.4 gives the solubility of air in water under an atmosphere with respect to temperature.

Temperature (°C)	s ($cm^3.L^{-1}$)
0	29.2
10	22.8
20	18.7
30	15.7

Table 4.4. *Solubility of air in water with respect to temperature*

4.6. References

Al-Khateeb, H.M. (2013). Pre-aeration and transversal baffling effect on aerated grit chamber performance of Kufa wastewater treatment plant. *International Journal of Science and Research (IJSR)*, 5(3), 1799–1803.

Albrecht, A.E. (1967). Aerated grit chamber design and operation. *Water Sewage Works*, 114(9), 95–106.

Andoh, R.Y.G. and Neumayer, A. (2009). Fine grit removal helps optimize membrane plants. *Water World*, 25(1), 28–32.

Annen, G.W. (1972). Efficiency of a grit chamber. *Water Research*, 6(4), 393–394.

Degrémont (2004). *Mémento technique de l'eau. Tomes 1 et 2*. Degrémont, Paris.

El Hifnawy, C. (2012). Conception de station d'épuration. Construction hydraulique. Thesis, Conservatoire national des arts et métiers, Paris.

Environmental Protection Agency (EPA) (2003). *Wastewater Technology Fact Sheet: Screening and Grit Removal*. EPA, Washington, D.C.

Finger, R.E. and Parrick, J. (1980). Optimization of grit removal at a wastewater treatment plant. *Journal of Water Pollution Control Federation*, 52(8), 2106–2116.

Gresch, M., Armbruster, M., Braun, D., Gujer, W. (2011). Effects of aeration patterns on the flow field in wastewater aeration tanks. *Water Research*, 45(2), 810–818.

Guerrée, H. (1967). *Pratique de l'assainissement des agglomérations urbaines et rurales*. Eyrolles, Paris.

Hoiberg, B. and Shah, M.T. (2021). CFD study of multiphase flow in aerated grit tank. *Journal of Water Processes Engineering*, 39, 115–128.

Kirschmer, O. (1926). Untersuchungen über den Gefällsverlust an Rechen : Mitteilungen des Hydraulischen Instituts der Technischen Hochschule München, Heft 1. PhD Thesis, Techniske Hochschule, Munich.

Lin, S.D. (2007). *Water and Wastewater Calculations Manual*. McGraw-Hill, New York.

Martin, E.J. and Martin, E.T. (1991). *Technologies for Small Water and Wastewater Systems*. Van Nostrand Reinhold, New York.

Maystre, Y. (1973). *Dessableur aéré*. Ecole polytechnique de Lausanne, Lausanne.

Morales, L. and Reinhart D. (1984). Full-scale evaluation of aerated grit chambers. *Journal of Water Pollution Control Federation*, 56(4), 337–343.

Neighbors, J.B. and Cooper, T.W. (1965). Design and operation criteria for aerated grit chambers. *Water Sewage Works*, 112, 12–19.

Patel, T., Gill, L., Faram, M.G. (2011). Grit removal from wastewater using secondary currents in open-channel flow around bends. *Journal of Environmental Engineering*, 137(11), 125–146.

Qasim, S. (1994). *Wastewater Treatment Plants: Planning, Design and Operation*. Technomic Publishing, Lancaster.

de Silva, V. and Bates, B.R. (2004). Fine grit removal challenges in wastewater treatment plants. *Proceedings of the Water Environment Federation*, 7, 258–264.

Water Environment Federation (1998). *Design of Municipal Wastewater Treatment Plants*. Water Environment Federation, Alexandria.

5

Primary Settling

5.1. General principles

Settling differs from grit removal inasmuch as it removes more than half of the suspended solids, whereas grit removal only removes sand and other mineral materials (heavier than organic matter).

The principle of primary settling was devised in an empirical manner rather than a rational one. The main reason was the lack of distinction between polluting loads and what the primary settlers were able to eliminate. For example, it is not uncommon to see a statement in many master plans of wastewater treatment plants or facilities such as: "Primary settlers are engineered to remove 60% of total suspended solids." No basis is given for such guarantees. In reality, it can be assumed that a 60% reduction is possible but not guaranteed. With a more advanced understanding of the various pollutants and removal mechanisms in structures, the most appropriate phrasing became: "Primary settlers are engineered to remove all total settleable suspended solids under operating conditions at the average flow rate in dry weather."

Since the total concentration of settleable suspended solids is a characteristic of wastewaters, a good design of the primary settler requires proper characterization of the wastewater.

The points discussed in this chapter include the physical and chemical mechanisms behind primary settling, performance (treatment with and without chemical reagents) and the factors affecting them, design concepts and design (configuration and depth), the impact of inflow rates, feed design, sludge collection and extraction, floating matter collection and recovery, and discharge of settled effluents.

Wastewater contains a wide variety of particles, the components of which vary in size, shape and density. The concentration of particulate matter can also vary considerably. The concentration of particles can be expressed in volumetric or gravimetric terms. The specific mass of the particles is determined by the density of the solid matter of the particles and the amount of water drawn in. Organic parts, in particular, usually carry large amounts of water.

In water process engineering, the concentration of particles is mainly expressed in $mg.L^{-1}$. However, in settling analysis and in the development of particle separation processes, the volumetric concentration ($ml \cdot l^{-1}$) is also important. When the volumetric concentration is high, the sedimentation rate of the individual particles is impeded by the upward motion of the water displaced by the downward motion of the suspended particles.

In wastewater, the settling of suspended solids is the displacement of these solid particles under the action of a gravitational acceleration field. It is the most commonly implemented liquid–solid separation method for the separation of suspended solids (TSS) and flocs during the liquid phase. Settleable products include suspended solids, flocculable colloidal materials and immiscible liquids of different densities, particles with a settling rate greater than 5 $m.h^{-1}$.

In order to have a good quality of settled water, effective sedimentation and correct sludge thickening, different horizontal zones of the unit are determined: clear water zone, sedimentation zone and thickening zone. The clear water zone is between 0.5 and 1 m high, the sedimentation zone between 1.5 and 2.5 m and finally the thickening zone between 0.5 and 1.5 m.

The parameters that are considered for designing and scaling the primary settlers are: settleable suspended solids, BOD_5 (ad2h), COD (ad2h). The settleable fractions are used to define the performance of the primary settlers and the produced primary sludge:

– soluble BOD_5 = particulate BOD_5 + total BOD_5 + colloidal BOD_5;

– soluble COD = particulate COD + total COD + colloidal COD;

– ad2h: after decantation 2 hours;

– BOD_5 ad2h = 60–67% of total BOD_5 (domestic effluent);

– COD ad2h = 60–67% of total COD (domestic effluent).

Reminder of the flow rates accounted for:

– Q_j: daily flow ($m^3.d^{-1}$);

– Q_{mts}: dry weather average flow;

– Q_{pts}: dry weather peak flow:

$$Q_{pts} = Q_j\, C_p/24$$

where $C_p = 1.5$ and Q_j is expressed in $l.s^{-1}$;

– Q_{nocts}: dry weather nocturnal flow.

There are many types of conventional settlers that differ, on the one hand, in the direction of water flow and, on the other, in the measures implemented for the disposal of sludge:

– cylindrical–conical settlers;

– horizontal settlers with sludge scraping;

– circular settlers with sludge scraping.

Scraping is carried out by way of a scraper bridge that drives one or more blades pushing the sludge towards one or more hoppers.

5.2. Conical or cylindrical–conical static settler

In this type of unit, the vertical component of the water velocity is predominant. These settlers are generally calculated for small facilities of 1,000–2,000 persons.

When the volume of sludge is low, and its density is high, these settlers can be scaled up to be larger.

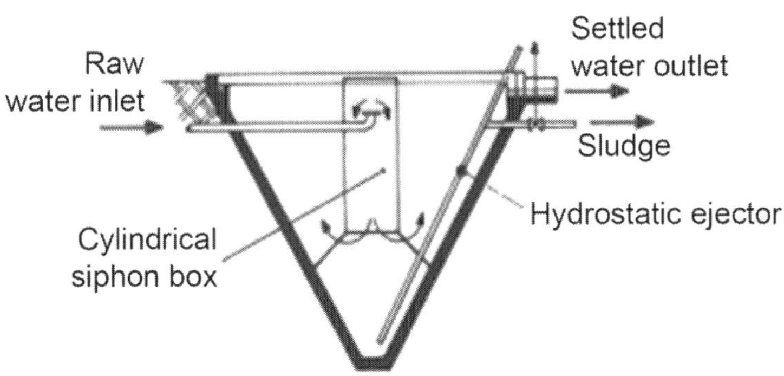

Figure 5.1. *Conical settler*

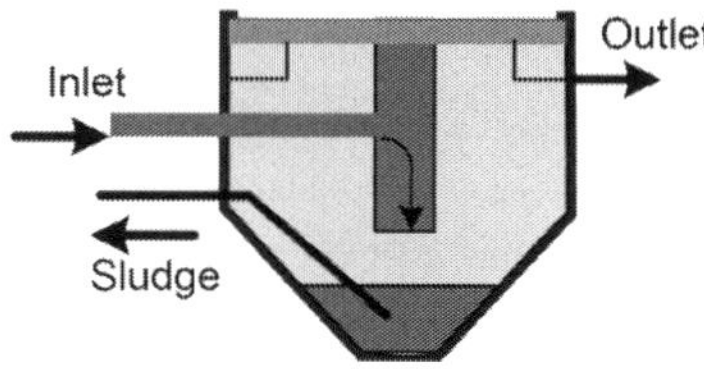

Figure 5.2. *Cylindrical–conical settler*

A 45–60° slope inside the conical section is adopted, which varies according to the nature of the water to be treated. The maximum surface load is 2.5 $m.h^{-1}$ at peak flow. At average flow, it varies between 0.6 and 1.2 $m.h^{-1}$. Knowledge of the velocity allows access to the area of the unit. Retention times vary between 1 h 30 min and 2 h with the average daytime flow.

Unscraped primary settlers with conical bottoms are often problematic because of various issues in pumping the sludge. Primary sludge is difficult to pump with small systems, which produces frequent blockages. Oversized pumping systems direct a large amount of clarified water to the sludge management system. This often leads to overloading the sludge management system and sludge return flow into the liquid chain. A septic tank or scraped settler is usually preferable.

5.3. Horizontal flow settler

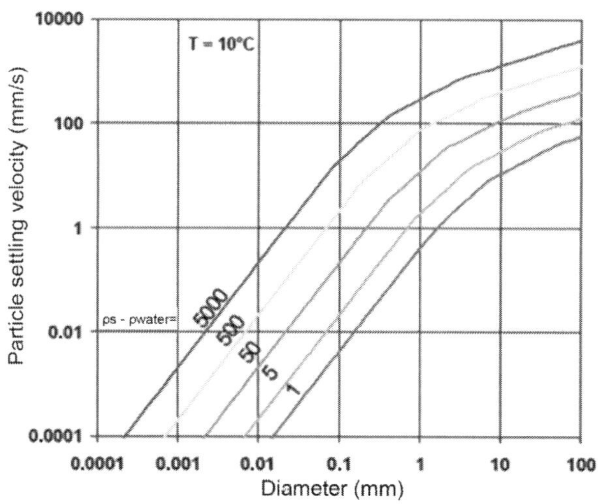

Figure 5.3. *Stokes velocities for particles of different diameters*

Depending on the process variant chosen for larger plants, or for those that occasionally have to receive high parasitic flows, it may be necessary to install a settler with automatic sludge and skimming extraction.

There are two types of rectangular settlers with scraping: scraper bridge settlers and chain settlers.

Settling is based on the theory of the Hazen load, according to which the smallest particle to be removed is the one that can be deposited along its path.

In a horizontal settler, for a length L and a water depth H, the smallest particle that will have time to settle is determined by the settling velocity (vs):

$$\mathrm{vs} = \mathrm{V}\frac{\mathrm{H}}{\mathrm{L}}$$

where V is the (horizontal) flow velocity in the settler.

5.3.1. *Dimensioning of horizontal settlers*

When dimensioning a horizontal settler (also called a "shallow settler"), the flocs entering the settler should be allowed to settle with enough velocity to access the bottom of the tank. The longitudinal schematic section of a tank (see Figure 5.4) shows the trajectories of the smallest particle and of an arbitrary particle with their respective falling velocities.

Therefore, for a floc to settle, it must reach the bottom of the settler before being removed, which is represented by the relation:

$$\mathrm{t} = \frac{\mathrm{L}}{\mathrm{v}} \geq \mathrm{tc} = \frac{\mathrm{H}}{\mathrm{vs}}$$

where t is the time of transit (h) in the settler, vs is the settling velocity of the flocs or particles ($m.h^{-1}$), H is the total height (m), L is the settler length (up to the outlet) (m), v is the flow velocity ($m.h^{-1}$) and tc is the settling velocity of the flocs (h).

Furthermore:

$$\mathrm{vs} \geq \frac{\mathrm{H.v}}{\mathrm{L}} = \frac{\mathrm{Q.H}}{\mathrm{L.H.l}} = \frac{\mathrm{Q}}{\mathrm{S}}$$

In theory, the particles will only be retained in the settler if vs is ≥ at v. We have:

$$vs = \frac{v.H.l}{l.L} = v\frac{H}{L}$$

Q is the treatment capacity ($m^3.h^{-1}$); v is the inlet velocity ($m.h^{-1}$); H, l and L are, respectively, the height (m), width (m) and length (m) of the settler; vs is the floc settling velocity ($m.h^{-1}$); and S is the area at the mirror of the settler (m^2). The above equation means that the treatment capacity is proportional to the mirror area of the settler. The smallest flocs should settle as long as their fall velocity is greater than or equal to Q/S. It is therefore important to control this parameter by obtaining the best flocculation possible that leads to flocs with the highest diameter possible and to an increase in the Stokes velocity.

In the event that the settling velocity of the flocs is lower than the transit or flow velocity (velocity at the mirror), the flocs end up at the end of the settler. Turbidity may be degraded (see Figure 5.4). When the floc settling velocity is greater than the flow velocity, the flocs settle very quickly at the beginning of the settler. This is the case for water with a high content of suspended solids.

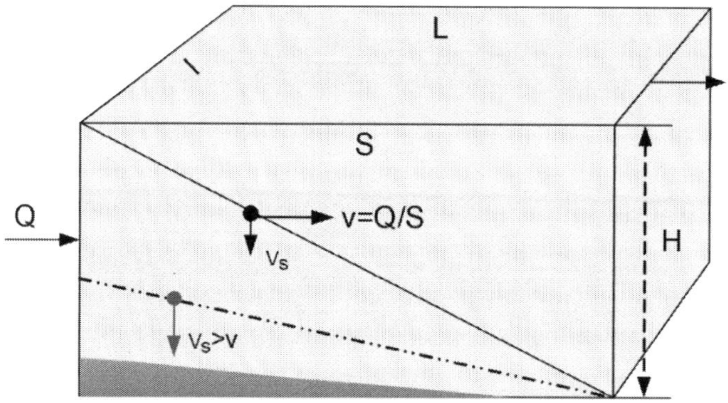

Figure 5.4. *Floc settling and flow velocity in a horizontal settler*

In Figure 5.6, trajectory II corresponds to that of any particle. vc is the settling velocity of a particle with a trajectory (I) and v's is the settling velocity of any arbitrary particle ($m.s^{-1}$). vh is the horizontal velocity of the water in the tank ($m.s^{-1}$).

When the settling velocity of the flocs is close to or equal to the flow velocity, the flocs settle along the water path and are dispersed over two-thirds of the length of the settler.

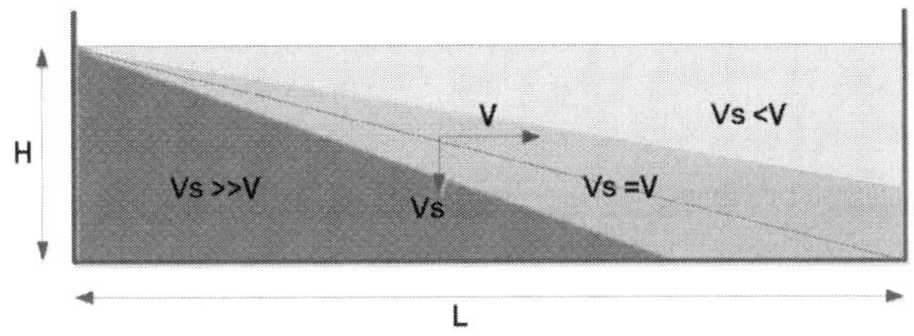

Figure 5.5. *Settling profile according to the settling velocity and the flow velocity*

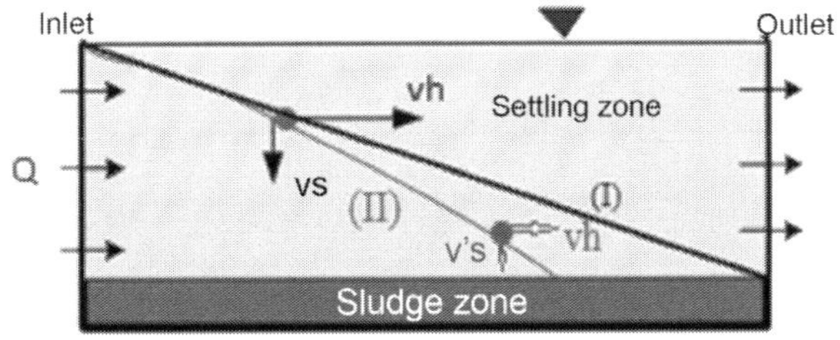

Figure 5.6. *Particle trajectory in a horizontal settler*

A horizontal settler is considered to be divided into four zones:

– an inlet zone, in which the flow is uniformly distributed following a horizontal direction and assimilated to a plug-flow piston;

– a settling zone, where the settling takes place with a liquid–solid separation that becomes more pronounced along the settler;

– an outlet zone, in which the flow converges towards the outlet weirs located at the surface of the settler;

– a sludge zone, where settled matter accumulates along the water path with a high concentration at the beginning of the settler. It is estimated that once a floc reaches the bottom of the settler, it is effectively removed from the water.

As a result, the settler length takes this into account such that the smallest settleable particle can be removed from the water. The length can extend up to 80 m while the maximal width observed is 20 m.

One of the main parameters to be considered is the mirror velocity ($m.h^{-1}$) because it is related to the velocity at which the settling flocs. Actually, the efficiency of the settling is theoretically independent of the tank depth, provided that the velocity at the mirror (flow velocity) is low enough so that the settled flocs are not put back into suspension by turbulence effect. This is the reason why we work with variable depths between 3.5 and 5 m as a precaution.

The mirror velocity ($m.h^{-1}$) is given by: $v = \frac{Q}{S}$. The retention time (h) is given by: $t = \frac{S.H}{Q}$. Q is the inflow ($m^3.h^{-1}$), S is the surface area at the mirror (m^2) and H is the water height (m).

Retention times in horizontal settlers vary between two and four hours, although some settlers exhibit retention times that can reach more than 12 hours.

The inflow velocity ($m.h^{-1}$) must be correctly chosen in order to avoid excessive dimensioning leading to resuspension of the settled flocs. This requirement affects the choice of length/width and height/width ratios for this type of settler. It is based on the relation:

$$vmax = \frac{Q}{H.l} = \frac{L}{t}$$

where l is the settler width (m). In practice, the L/l ratio varies between 3 and 5 with a value of 4 recommended.

The performance of horizontal settlers is sensitive to temperature differences because of the low flow velocity (< 0.5 $m.h^{-1}$) which characterizes these facilities, particularly in periods of hot temperatures. A difference in temperature of only 1°C is sometimes enough to cause a decrease in efficiency explained by changes in density.

The suspended particle is subjected to a horizontal velocity v related to the inflow rate and also to the settling velocity vs. In this type of unit, it is necessary to provide an inlet zone, an outlet zone and a sludge zone.

The longitudinal schematic section of a tank helps highlight the paths of the smallest particle and that of any particle, with their respective velocities (see Figures 5.4 and 5.5).

If l designates the tank width (m), and Q the feed water flow ($m^3.s^{-1}$), the velocity of water flow is equal to:

$$v = Q/(H.l)$$

where H is the height of the tank (m).

According to Figure 5.5:

$$\frac{\mathrm{v}}{\mathrm{vs}} = \frac{\mathrm{L}}{\mathrm{H}} = \frac{\mathrm{Q}}{\mathrm{H\,l\,vs}}$$

$$\frac{\mathrm{Q}}{\mathrm{vs}} = \frac{\mathrm{H\,L\,l}}{\mathrm{H}} = \mathrm{l\,L} = \mathrm{S}$$

(S is expressed in m²).

This gives the area of the tank (S). This equation shows that the tank height has no impact on sedimentation. At the theoretical value S, the corresponding areas of the inlet and outlet zones ($\Delta\ L_1x_l$ and $\Delta\ L_2x_l$) should be added.

We take a sludge height equal to H = 0.5 m.

When the particles do not have the same particle size, the settling velocities are different. For a given diameter, there is a fraction of solids with a settling velocity vs = Q/S, which is called x_0, and another fraction called (1 - x_0), whose settling velocity is greater than vs. The settling percentage is given by:

$$\text{settling \%} = (1 - x_0)\int_0^{xo} \frac{vs}{v's}\,dx$$

$1 - x_0$ is the fraction of particles with velocities greater than v's.

$\int_0^{xo} \frac{vs}{v's}\,dx$ is the fraction of the particles with velocities smaller than v's.

EXAMPLE.– The goal is to remove all particles with a 0.08 mm diameter from wastewater. The water flow rate is 720 $m^3.h^{-1}$, and the specific mass of suspended solids is equal to 1,200 $kg.m^{-3}$.

Preliminary tests have shown that the settling velocity of these particles is:

$$vc = 0.05\ cm.s^{-1} = 1.8\ m.h^{-1}$$

The calculation using Stokes's law gives, with a Poiseuille viscosity 10.10^{-4}:

$$vs = \frac{9.81\ (1{,}200 - 1{,}000)}{18\ 10^{-3}}(0.08\ 10^{-3})^2 = 6.7\ 10^{-4} m.s^{-1}$$

$$vs = 2.5\ m.h^{-1}$$

What is the area of the tank for removing 100% and then 75% of the solids?

S = Q/vs

For 100%, we obtain: S = 720/2.5 = 288 m^2.

For 75%: S'= 0.75 S = 216 m^2.

Static horizontal flow settlers are of very old units, systematically requiring that sludge be drained by the total evacuation of the water from the basin. The best settling efficiencies are obtained in the laminar regime. The bottom of the tank is gently inclined with an angle ranging from 5 to 60°. Retention times are about 1 hour 30 minutes to 2 hours at daytime flow.

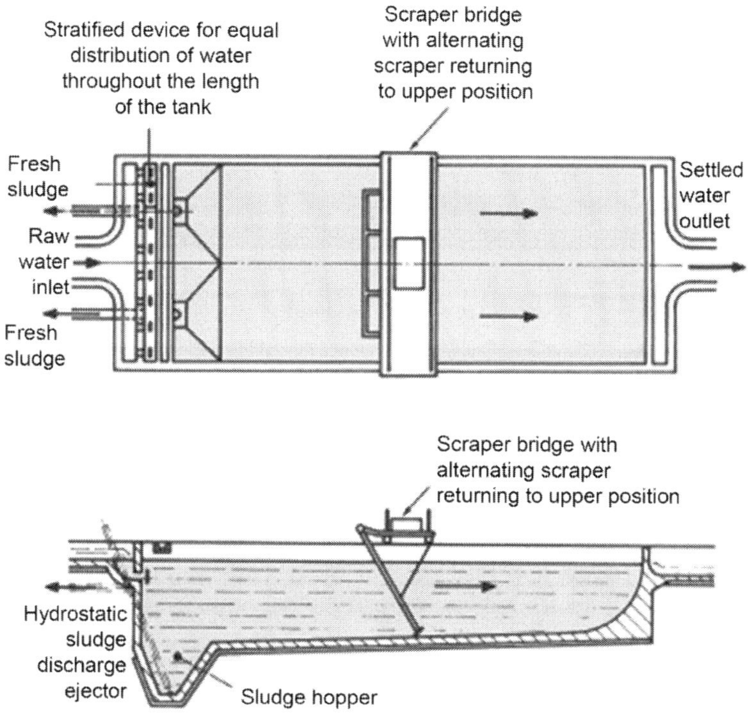

Figure 5.7. *Rectangular chain settler*

Modern rectangular settlers have an inlet structure located at one end of the rectangular tank and the outlet troughs are located at the opposite end. The gutters at the outlet end can be designed as fingers pointing to the tank or as one or more gutters extending through the tank. An additional weir mounted directly on the back wall can be added to the troughs to complete them.

The scraper cranes move in a back-and-forth system before the scraping in a countercurrent movement. During the scraping period, the velocity may not exceed 3 $cm.s^{-1}$, which limits the scraping devices to 60 m. A 3–6 length/width ratio is adopted in these facilities.

A 2.5–4 m height is often used, and the bottom slope is 1%. Chain settlers allow for a continuous scraping of the sludge and floating bodies by a series of scrapers mounted between two parallel endless chains and rotating along the vertical walls of the tank. The big disadvantage of these devices is the rapid wear of all the submerged moving parts.

The sludge pits are located at the raw water inlet, and skim collection is done immediately upstream of the outlet of the settled water.

The settled sludge is scraped to the sludge hopper by means of transverse scrapers mounted on an overhead crane or, less wisely, by metal scrapers. Chain scrapers are only recommended after fine sieves (maximum openings of 6 mm), because otherwise debris accumulates quite quickly on the chains.

5.4. Circular settlers

Figure 5.8 shows a diametrical section of a circular tank.

By calling t the retention time of the effluent in the tank, we can write that:

t = tank volume/inlet flow

The settling velocity vs is equal to: vs = H/t, hence vs = H.Q/volume = Q/S.

The inlet structure is located in the center of the tank. The treated water usually leaves the tank through an outlet chute positioned along the tank periphery and equipped with one or two overflow weirs. The distance between the wall and the outer weir must be at least 0.40 m.

The scraper is attached to a framework rotating around the tank axis. Its design is different depending on the models manufactured. It can feature a single spiral-shaped blade or a series of squeegees disposed as a jalousie. It consists of a hollow central shaft that receives the raw water from which it is generally distributed by an annular siphonic partition.

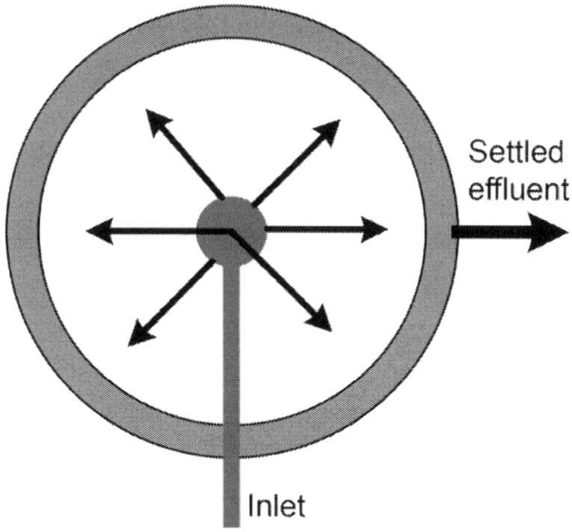

Figure 5.8. *Top view of a circular settler*

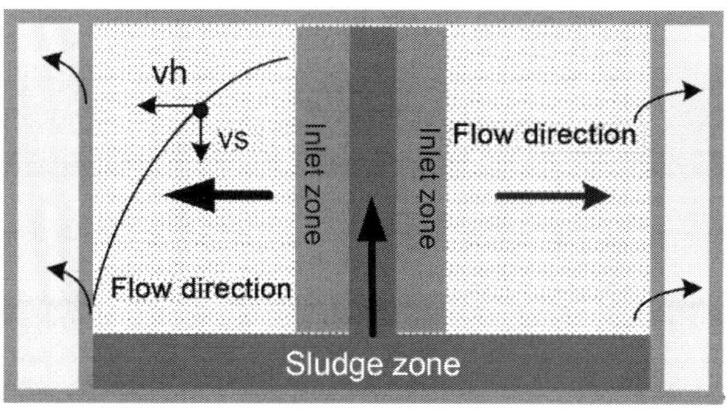

Figure 5.9. *Diametrical tank cross-section*

The scraping device allows the sludge to be conveyed towards a central hopper from which the extraction pipes start. In addition, a surface scraper drives the floating matters to a skimmer from where they can be sent either to the sludge hopper or to a separate plant. The most common construction is the use of the peripheral drive scraper bridge. The surface scraper is rigidly attached to the rotating deck. When the model is engineered with a central drive, the framework consists of two radial arms suspended from a central ring gear. In all cases, a geared motor unit ensures the rotation of the system. The bottom slope on which the sludge is scraped has a 4–10% slope.

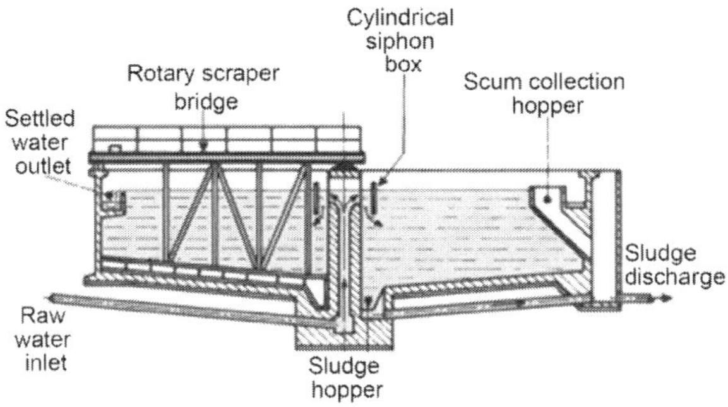

Figure 5.10. *Circular settler with sludge scraping*

The average height of scraped circular settlers is 3–3.5 m. The diameter can reach 90 m. However, they become sensitive to wind effects, which can disrupt the proper distribution of the liquid.

The treated water can also leave the tank through weirs mounted directly on the outer wall. The outlet trough is then positioned on the outer side of the tank and can be used as a track for the overhead crane.

The settled sludge is scraped up to the hopper by way of a helical lower scraper. The scraper is driven around the tank by a bridge equipped with a periphery-mounted geared motor. The bridge must be continuously operated if pre-precipitation is used.

The general criteria for designing a conventional primary settler are summarized in Table 5.1.

Parameters	Unit	Range	Typical value
Primary settler followed by secondary biological treatment			
Retention time	h	1.5–2.5	2
Upflow velocity			
At average Q	$m^3.m^{-2}.h^{-1}$	1.25–2.50	1.65
At peak Q	$m^3.m^{-2}.h^{-1}$	3.4–5	4.2
Overflow rate	$m^3.m^{-2}.h^{-1}$	5.2–20	11

Table 5.1. *Design of a conventional mechanized primary settler*

The retention times shown in Table 5.1 should be adapted according to temperature (T) according to the following formula (Metcalf and Eddy, Inc. 2014):

$$\text{Correction factor} = 1.82\ e^{-0.03T}$$

A primary settling system with continuous sludge scraping can remove up to 70% of SS and 30%–40% of BOD_5 (Metcalf and Eddy, Inc. 2003), but efficiency decreases with reduced retention time. Moreover, these estimates are based on data sources that are highly variable, and should be viewed with caution (see Figure 5.11).

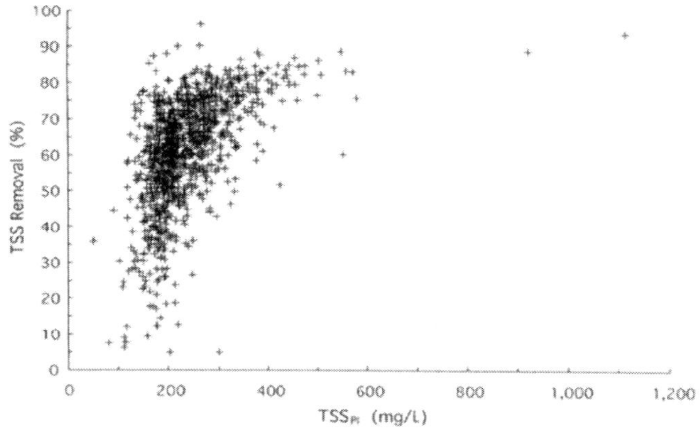

Figure 5.11. *SS removal percentage based on the concentration of SS in raw water (Clarifier Design, 2005)*

It can be observed that efficiency increases when the SS concentration in the raw wastewater increases.

Primary sludge is highly valuable from an energy point of view. These can be directed to an anaerobic digester to produce natural gas (methane) of cyclic origin.

5.4.1. *Example design of a primary settler*

We want to compute the dimensions of a primary settler of an urban water treatment plant that receives, in about 15 hours, 9,000 m^3/d of effluent with a suspended solid content that must be reduced from 380 ($mg.L^{-1}$) to 120 ($mg.L^{-1}$).

For that, a settling experiment is performed using a cylindrical column with a minimum diameter of 20 cm and a height of 2.25 m, with samples taken at 60, 120 and 180 cm from the free surface.

After sufficient homogenization, samples are collected at 5, 10, 20, 40, 60, 90, 120 min at the three sampling points, on which suspended solids are determined.

The percentage of solid removal for each sample is calculated, and the results are plotted on graph paper, with the observation times on the horizontal axis and the dimensions of the sampling points on the y-axis. From these elements, the curves of equal elimination can be drawn.

For different settling points, the average removal efficiencies are calculated by weighting the partial efficiencies observed relatively to the different water slices. The results obtained are then used to draw the total removal curve (in %) as a function of retention time. The minimum retention time t is then found to guarantee the required removal efficiency.

Consequently, the upward velocity is easily deduced, which must be adapted when a useful settling height will be set:

$$vs = H/t$$

The initial SS concentration is 380 mg.L^{-1}.

The experimental results obtained from monitoring the residual SS for different settling times and different heights are presented in Table 5.2.

	Time (min)						
	5	**10**	**20**	**40**	**60**	**90**	**120**
60	225	170	150	135	105	100	90
120	310	250	210	160	145	115	100
180	330	260	235	175	155	140	110

Table 5.2. *Experimental results for an initial SS content of 380 mg.L^{-1} (data in mg.L^{-1})*

Next, the concentrations are then calculated (Table 5.3), for which the removal efficiencies of the SS achieved are 40, 45, 50, 55, 60, 65, 70 and 75%.

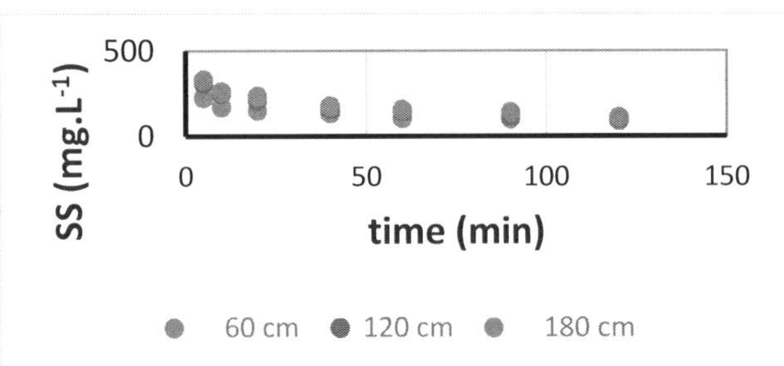

Figure 5.12. *Evolution of settled SS according to time and height. For a color version of this figure, see www.iste.co.uk/gaid/wastewaterengineering1.zip*

Removal efficiencies (%)	40	45	50	55	60	65	70	75
SS concentration (mg.L^{-1})	228	209	190	171	152	133	114	95

Table 5.3. *Removal efficiencies and SS concentrations*

The SS curves experimentally obtained according to time are then drawn up for the three depths (Table 5.4).

Height (cm)	Time (min)						
	5	**10**	**20**	**40**	**60**	**90**	**120**
60	41	55	60	64	72	74	76
120	18	34	45	58	62	70	74
180	13	32	38	54	59	62	71

Table 5.4. *Observed efficiencies for different heights and different settling times (in percent)*

The efficiency curves (see Figure 5.13) are obtained from the values in Table 5.4.

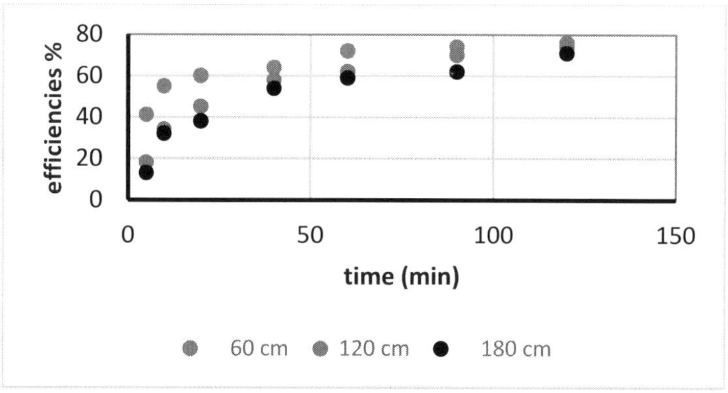

Figure 5.13. *Efficiency curves for different heights. For a color version of this figure, see www.iste.co.uk/gaid/wastewaterengineering1.zip*

Based on the efficiency curves, and following the principle of linear interpolation between two curves, the average efficiencies calculated are obtained by the relation:

$$\text{Average efficiency} = \sum_{0}^{150} \Delta hij \frac{(Rhi + Rhj)}{\sum hij}$$

The results are presented in Table 5.5.

Time (min)	15	20	30	45	60	75	90	120
Average efficiencies (%)	53.79	58.13	62.70	65.47	71.52	73.66	75.34	79.33

Table 5.5. *Average efficiencies according to time*

The required retention time to reduce the concentration from 380 mg.L^{-1} to 120 mg.L^{-1} can then be determined.

To shift from 380 mg.L^{-1} to 120 mg.L^{-1}, an efficiency of 68.4% must be obtained, which is equivalent to a retention time of 50 minutes.

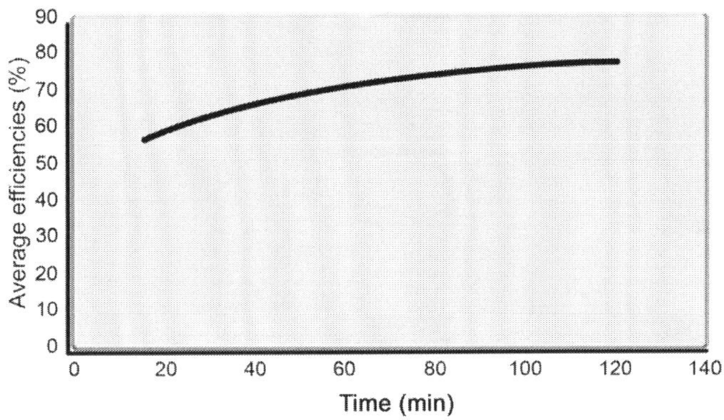

Figure 5.14. *Average efficiencies (%) according to time (min)*

5.5. Design of primary mechanized settlers

The settler can have a circular or rectangular shape.

5.5.1. *Rectangular settler*

It is in the central zone, which is the theoretical zone where liquid currents are considered to be parallel and horizontal, that the particles must be collected.

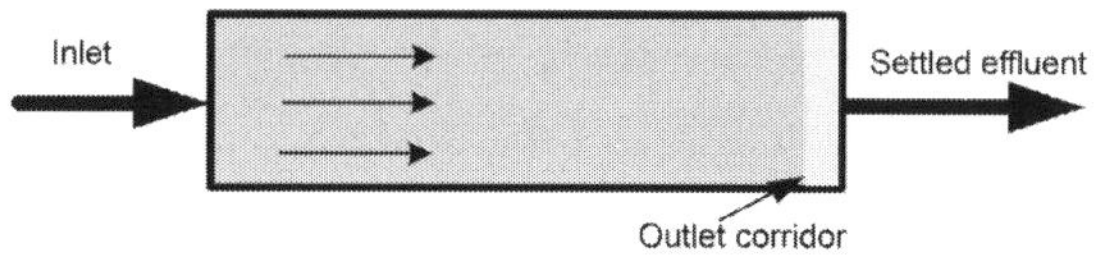

Figure 5.15. *Horizontal primary settler traversal flows*

If vs_l denotes the limit settling velocity of the particles and v is the velocity of the water current, it can be written that:

$$\frac{v}{vs_l} = \frac{L}{h}$$

Since $v = \frac{Q}{hl}$, it is then obtained that: $vs_l = v\,\frac{h}{l} = \frac{Qh}{Lhl} = \frac{Q}{S}$.

5.5.2. *Circular settler*

Let ro denote the radius of the central inlet pipe, r that of the settler of height H, and vo and v_h the water horizontal velocities at the inlet and at the periphery of the settler. If we consider a particle undergoing sedimentation, its horizontal velocity is that of the water at the point in question, at a distance r from the axis of the device:

$$vh = dr/dt = vo\ ro/r$$

Its vertical velocity is constant, that is:

$$vv = dz/dt$$

By integrating these equations, we have:

$$2\ vo\ ro\ t = (r^2 - r^2o) \text{ and } z - zo = vh\ t$$

It can be written that the flow rate introduced into the device is:

$$Q = 2\ \pi\ ro\ h\ vo$$

which becomes, by replacing vo by its value:

$$Q = 2\ \pi\ ro\ h\ vo(\frac{(r^2\ - r^2o}{2rot})$$

or:

$t = \pi h\,(r^2 - r^2o)/Q$

which we carry forward into the equation of the trajectory:

$$z - zo = \pi h\, v \frac{r^2 - r^2o}{Q}$$

For an inlet flow of 9,000 $m^3.d^{-1}$ in 15 hours, that is, Qm = 166 $l.s^{-1}$, the peak coefficient Cp can already be calculated by the relation:

$$Cp = 1.5\, \frac{2.5}{\sqrt{Qm}} = 1.7$$

The peak flow is: Qp = 0.166 x 3,600 × 1.7 = 1,020 $m^3.h^{-1}$.

Given that the SS average removal efficiency is obtained in 50 min, or 0.833 h, the limit upflow velocity (v up) is, for a depth of 2 m, equal to:

limit vup = 2/0.833 = 2.4 $m.h^{-1}$

The settler area is then:

S = Qp/vup = 1,020/2.4 = 425 m^2

whether the settler is rectangular or circular.

For a circular settler, a radius of 12 m can be expected; increased by a length of 2 m for the acceleration zone, a radius of 14 m leads to a length of 88 m.

For a rectangular settler, in order to respect a velocity of the overflowing wave of 10 $m.h^{-1}$ at most, the minimum length required is 60 m with the dry weather flow and 100 m at the peak flow rate. The width would then be a little more than 4 m.

5.6. Performance of conventional primary settlers

Sludge thickening in primary settlers is more detrimental to the wastewater treatment system (e.g. in the form of reduced settling of activated sludge), resulting in increased organic loads and the production of hydrogen sulfide and volatile acids, than advantageous for the sludge treatment system.

To simplify, a primary settler that is perfectly designed and operated will have a total concentration of suspended solids (TSS) in the effluent equal to the concentration of non-settleable SS in raw wastewater. Depending on the inlet flow rate (average flow, peak flow, maximum flow), there will be a change in settleable

SS concentrations with an increase in the biochemical oxygen demand (BOD) of settleable particles and in the chemical oxygen demand (COD).

As a unitary operation, physical forces predominate in SS removal in primary settlers. It is perhaps for this reason that many consider a primary settler to be a constant percentage process of SS removal. The objective of the primary settler process is to remove settleable SS, whether these solids already exist in the raw wastewater or whether they consist of precipitated solids generated as a result of chemical addition for improved removal of SS, phosphorus or heavy metals.

Although primary settlers only remove settleable SS, their performance has always been quantified on the basis of the total suspended solid removal efficiency calculated from the relation:

% removal efficiency = 1 – (SSi/SSs)

where:

– SSi = SS in raw water ($mg.L^{-1}$);

– SSs = SS in settled water ($mg.L^{-1}$).

By removing settleable suspended solids, primary settlers remove some of the BOD, COD and phosphates associated with these particles.

Since the biological processes used downstream are dimensioned according to the quantity of biodegradable material present in the primary effluent, the performance of primary settlers can be quantified according to COD or BOD following the same formula as previously:

% removal efficiency = 1 – (CODi/CODs)

where:

– CODi = COD in raw water ($mg.L^{-1}$);

– CODs = COD in settled water ($mg.L^{-1}$).

The same is applied for BOD.

The fundamental principles of primary settler design are based on settling analyses of free (type 1) and diffuse (type 2) particles. This basis has been used for most primary settler designs operating to date. We have:

– V = settler volume (m^3);

– Qj = daily flow rate of raw water ($m^3.d^{-1}$).

The volume is obtained by the relation:

$$V = S\,H$$

where:

– S: primary settler area (m^2);

– H: water height in the settler (m).

Hydraulic load ($m^3.m^{-2}.d^{-1}$) is equal to:

$$C_h = Q/S$$

The following relation is proposed to estimate the SS removal efficiency:

$$\%\text{ removed SS} = [1 - (SSnd/\,SSi)] - [1 - (SSnd/\,SSi)]\,e^{-\lambda/Ch}$$

where λ is the constant related to the nature of SS in the raw water.

The important factors affecting the performance of the primary settler can be summarized as follows:

– concentration of non-settleable SS (SSnd);

– concentration of SS in raw wastewater (SSi);

– settling characteristics of settleable SS;

– hydraulic load (Ch);

– ratio of CODp (or BOD_5p)/SS in raw wastewater.

A good primary settler design therefore starts with a good knowledge of the characteristics of wastewater. The removal percentage depends on the COD/SS ratio and SS removal for an inflow COD/SS ratio < 3.0.

SS reduction (%)	% COD reduction
0–65	65 × inlet COD/SS ratio
65–75	70 × inlet COD/SS ratio
75–100	75 × inlet COD/SS ratio

Table 5.6. *COD reduction percentage according to SS reduction (theoretical approach)*

The reduction of the other parameters, BOD, N and P, also depends on the corresponding ratios, namely BOD/SS, total N/TSS and total P/SS and SS removal efficiencies:

– For a total N/SS ratio < 0.25, the reduction percentage in primary settling would be 14%.

– For a total P/SS ratio < 0.06, the reduction percentage in primary settling would be as presented in Table 5.7.

SS reduction (%)	Percentage of total P reduction in primary settling
0–60	35 × total P/SS ratio
60–70	40 × total P/SS ratio
70–80	45 × total P/SS ratio
80–100	50 × total P/SS ratio

Table 5.7. *Percentage reduction of suspended solids as a function of the ratio total P/SS*

5.6.1. *Settler depth*

Circular settlers		Rectangular settlers	
Tank diameter (m)	**Minimal water level (m)**	**Tank width (m)**	**Minimal water level (m)**
19	3	6	3.5
20–29	3.5	7–8	4
29–30	4	9–10	4.5
40	4.5	–	–

Table 5.8. *Recommended dimensions for conventional settlers*

– Recommended maximal water level for all units: 5 m.

– Rectangular tank length:

- with transverse scraper: maximum 60 m;
- with mobile scraper (trac-vac): maximum 45 m.

– L/l ratio:

- average: 5 < L/l < 7;
- optimum: L/l = 6.

5.7. Lamellar settlers

The removal of solid matter from raw wastewater by the primary settlers is carried out according to the principle of gravitational separation. Since the existing solids vary considerably in size, shape and density, gravitational separation should theoretically take into account the distribution of sedimentation velocities of all the different solids. However, SS from raw wastewater are categorized as settleable or non-settleable. Similarly, total COD (or BOD) is soluble particulate associated with settleable SS or with non-settled SS. Non-settleable COD is composed of soluble COD and particulate COD associated with non-settleable SS. The goal of primary treatment associated with chemical reagents is to produce a settled effluent with SS and COD contents lower than non-settled SS and COD levels, measured without the addition of chemicals.

The first papers published by Hazen (1904) and Camp (1945) developed the theory of sedimentation tank design. They were the first to establish that the disposal of SS in gravity clarifiers depends only on the area and not on the depth of the tank. Plates or tubes installed at an angle in a clarifier will significantly increase the settling area available in a given footprint.

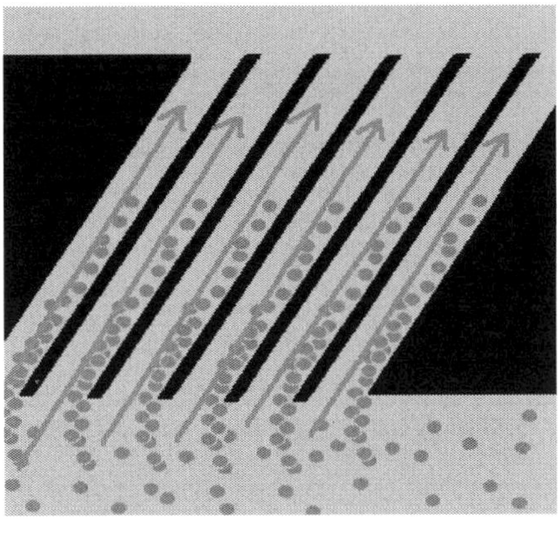

Figure 5.16. *Schematic diagram of a sludge flow over a plate*

These tanks have a more modern design and are based on the fact that the only fundamental quantity is the horizontal area Sh of the unit. Indeed, a particle is captured if its settling velocity is greater than the Hazen velocity (or surface hydraulic load). Particle retention is therefore theoretically independent of the height of the structure.

5.7.1. *Theory and general principles*

There are two types of conventional lamellar settlers that differ according to the direction of water flow.

When the water flows (downflow) in the same direction as the flocs, the settlers are co-current. When the water flows in the opposite direction of the flocs, the settlers are countercurrent:

– Co-current systems facilitate floc removal, but short circuits are often observed.

– In countercurrent systems, the feed of flocculated water is carried out below the lamellar block. This results in a mix between the sludge that slides along the plates and migrates towards the bottom of the settler. The water is evacuated from the top and then is recovered by chutes. Clarified water is obtained in the part where the floc concentration is lowest, resulting in better clarified water. In addition, countercurrent operation implies a simpler design than the co-current mode. The control of the upward hydraulic flow is easier than that of the downward flow of co-current systems. This facility compensates for the difference in the projected area between these two systems.

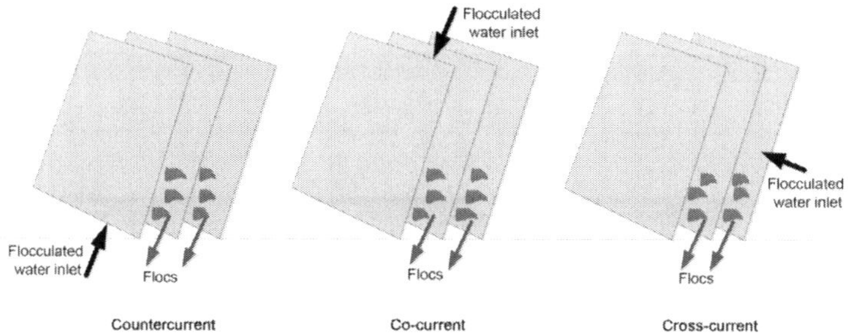

Figure 5.17. *Different lamella block feeding systems*

There are also cross-flow lamella settlers in which the floc flow is perpendicular to the direction of the water. The countercurrent design is often preferred because of its simpler design.

Lamella settlers are divided into several categories:

– with or without solid contact;

– with or without weighted flocculation.

If the same theory used for horizontal settlers is applied to lamella settlers, the parameter relating to the presence of lamellae must nevertheless be taken into account. A lamella module designates a device comprising lamella inclined at an angle, either flat or corrugated and stacked at a short distance from each other.

A lamella settler is considered to be the compilation of a large number of elementary settlers, each of which consists of the defined volume between two settling plates.

Therefore, two back-to-back plates form a module composed of hexagonal sections. The flocculated water is distributed under the modules through an orifice whose particularity is to extend over the whole width of the device. Water upflows between the plates in the opposite direction (against the current) to the flocs, which settle on the lamella and slide downward due to gravity.

The existing space between each lamella represents a modular settler whose Hazen velocity (linked to the cut-off threshold) is characterized by the ratio between the inlet flow and the projected area of the lamella. It is expressed by the following relation:

$$v\ (m.h^{-1}) = Q/Sl \cos \alpha$$

where Q is the flow rate ($m^3.h^{-1}$), Sl is the area of the lamella (m^2), and α is the angle of inclination of the lamella with respect to the horizontal.

The settler depth has no influence on the cutoff capacity provided that the deposited material is discharged as it accumulates and does not disrupt the inlet water flow.

The combined use of a high mirror speed in combination with inclined plates (45–60°) allows for continuous self-scraping. There is very little floc accumulation at the plates. Let us define a coordinate system Ox, Oy, Oz, Ow where:

– Ox is parallel to the lamellae;

– Oy is perpendicular to the lamellae;

– Oz is perpendicular to the xOy lamella;

– Ow is perpendicular to xOy and is related to the settling direction of the flocs on the lamella.

Then, the water moves following Ox (clarified water effluent) with a flow velocity between two lamella (see Figure 5.18) and the flocs are subjected to a Hazen velocity (vertically following Ow) corresponding to the direction of settling and deposit of the flocs. The angle α is therefore determined such that the settled floc can slide to the bottom without accumulating at a given location on the lamellas. This is why the total area is calculated so that the laminar regime (Reynolds < 500 and Froude numbers > 10^{-5}) is such that the gravitational forces allow the flocs to slide down to the bottom of the settler.

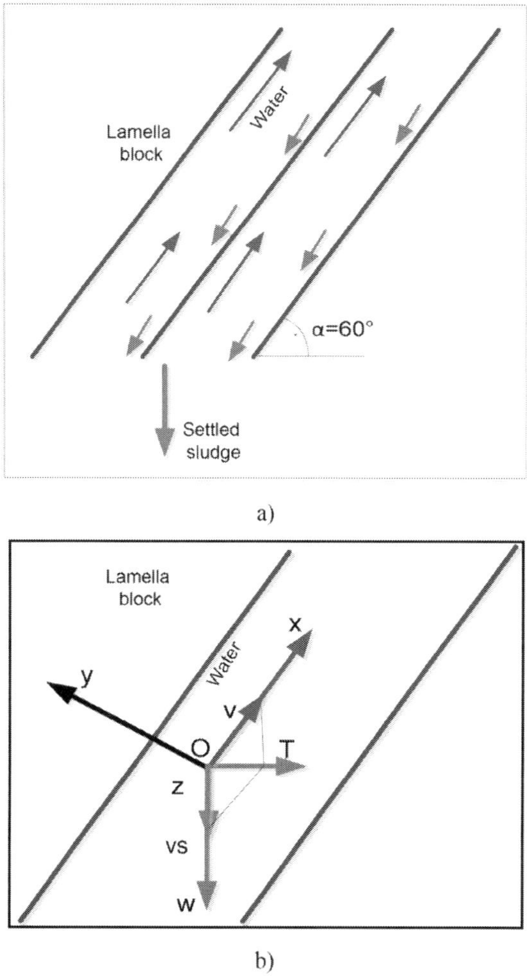

Figure 5.18. *a) Countercurrent operation principle; b) liquid–solid separation spatial representation*

Figure 5.19 shows floc behavior between two lamellae. When a particle moves along the lamellae, the necessary time for the particle to move through the lamellae is $L/(V-v_s \sin\alpha)$. The time required to settle on the lower lamella is $H/v_s \cos\alpha$.

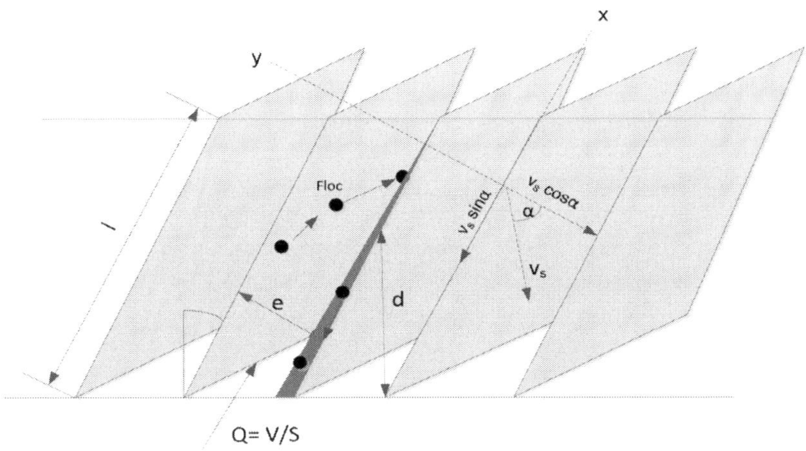

Figure 5.19. *Floc path on lamellar plates*

COMMENT ON FIGURE 5.19.– *α is the inclination angle of the plates (or lamellae), V is the mirror velocity at the mirror (m.h^{-1}), S is the settler area (m^2), Q is the transverse flow rate (m^3.h^{-1}), e is the spacing between the lamellae (m), d is the vertical distance between two lamellae (m), v_s is the settling velocity of the flocs (m.h^{-1}), H is the plate length (m) and l is the projected length of a plate (m).*

Therefore, the smallest particle that can be removed by the lamellar settler is determined by the settling velocity:

$$vs = v\frac{H}{L\cos\alpha + H\sin\alpha}$$

If $\alpha = 0$, we find $v_s = v$ H/L relatively to the horizontal settlers.

If $\alpha = 90°$, $v_s = v$. In this case, the settling velocity of the flocs must be greater than the flow velocity of the water through the lamellae for the settling to occur. It is only if $v_s < v$ that the particles are removed.

5.7.2. *Dimensioning basics for lamellar settlers*

Different processes are applied in which the water and mud flow in different directions. Countercurrent systems where mud and water flow in opposite directions are the most commonly used flow model (Dudley 1994). In countercurrent systems, the water inflow is achieved below the lamella block. The flow is upwards in the spaces between the lamellae, and the influent is generally fed longitudinally through the inlet ports located under the lamella to ensure better flow distribution.

As a consequence, the sludges that slide along the lamella mix together and migrate towards the bottom of the settler. The water is evacuated from the top and then is recovered in the pipes. The principle of the lamella settling consists of using lamellae close to each other to increase the available settling area in a given tank volume. In order to ensure that the sludge is discharged, the lamellae are tilted at an angle α with respect to the horizontal. The area available for settling then corresponds to the total projected surface (TPS) on the horizontal of all the plates:

$$TPS = n\ Sp\ \cos\alpha$$

where n is the number of lamellae, Sp (m^2) is the unit area of each lamella and α is the angle of inclination of the lamella with respect to the horizontal (45°–60°).

Particle removal (flocs, suspended solids) in lamellar settlers depends on the developed area but not on the depth of the settler (Hazen, Camp). This is what is observed, for example, when a tube containing flocs is inclined, compared to the same tube in a vertical position:

$$\text{Total projected surface (TPS , mm}^2) = \text{settling area} \times \cos\alpha$$

Therefrom, we deduce the TPS/mirror surface ratio. Sm is the mirror surface. It is equivalent to the ratio between the mirror velocity (vm, $m.h^{-1}$) and the Hazen velocity (v_H, $m.h^{-1}$):

$$\frac{TPS}{Sm} = \frac{Vm}{VH}$$

Particle removal (flocs, suspended solids) in lamellar settlers equipped with flat plates is presented in Table 5.9.

Parameters	Equations
Hazen velocity ($m.h^{-1}$)	$vH = Q/Sp$
Lamella projected surface (m^2)	$Sp = Lp\ lp.\cos\alpha$
Total projected surface (m^2)	$TPS = n.Lp.lp.\cos\alpha$
Mirror velocity	$V = Q/S$
Inclination angle	$\alpha = 45–60°$
Plate projected length	$L = H/\sin\alpha$

Table 5.9. *Dimensioning parameters for flat lamellar plates (Lp: plate length (m); lp: plate width (m); n: number of plates)*

Therefore, two parameters are taken into consideration when dimensioning lamella settlers:

– Hazen velocity ($m.h^{-1}$), given by: $vH\left(\frac{m}{h}\right) = \frac{Q}{TPS} = \frac{Q}{n.Sp.\cos\alpha}$.

– Mirror velocity ($m.h^{-1}$), given by: $v\left(\frac{m}{h}\right) = \frac{Q}{S}$.

S is the mirror surface of the settler (m^2), Sp is the lamella surface (length × width in m^2), TPS is the total projected surface (m^2), e is the distance between two lamellae (m), L is the settler length (m), n is the number of lamella and Q is the inlet flow rate ($m^3.h^{-1}$).

It is possible to calculate vH using (n – 1) lamellae in order not to count the one at the end of the settler that is not effective. However, the overall calculation remains the same, especially for high-throughput plants.

The number of lamellae n is equal to:

$$n = \frac{L \sin\alpha}{e}$$

For an accurate computation of L, only the effective length containing the lamellae should be counted, because due to their inclination, a small area at the end of the settler is not covered.

As with hexagonal cell lamellae, there is a specific ratio between total projected surface (TPS) and mirror surface (Sm) which is involved in the efficiency of floc separation.

We have:

– Lp: lamella length (mm);

– lp: lamella width (mm);

– e: distance between two lamellae (mm);

– e.lp: flow area (mm^2);

– e.lp/sinα: mirror area between two lamellae (Sm, mm^2);

– Lp.Lp: settling surface (mm^2);

– settling surface × cosα: total projected surface (TPS, mm^2).

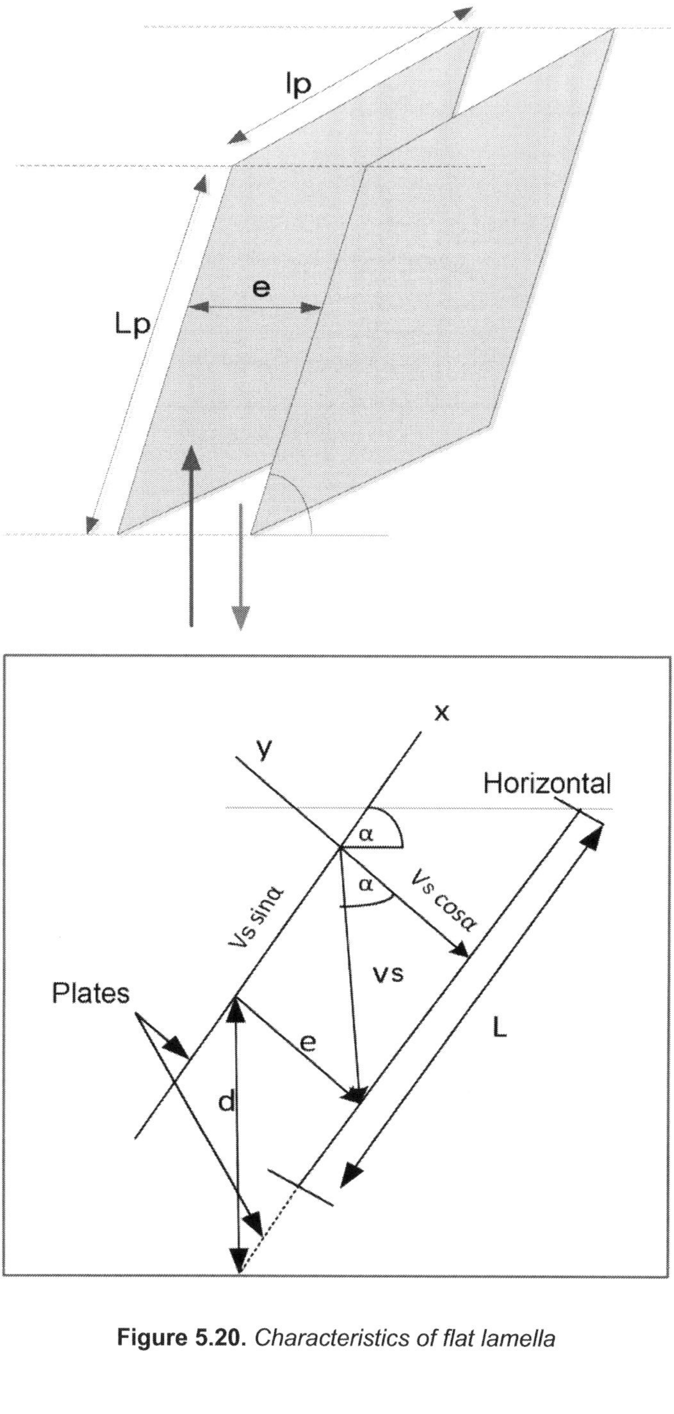

Figure 5.20. *Characteristics of flat lamella*

The TPS/mirror surface ratio is equivalent to the ratio between the mirror velocity (vm, $m.h^{-1}$) and the Hazen velocity (vH, $m.h^{-1}$):

$$\frac{TPS}{Sm} = \frac{vm}{vH}$$

The lamella length varies between 1.5 and 2 m. The total projected surface is the surface of a lamella multiplied by the number of lamellae and by cosα. The spacing is about 50 mm and can be as high as 80 mm.

This allows for an increase in the effective settling surface by a factor of 6–12, thus enabling a higher peak flow rate to be processed. Reducing the spacing between lamellae results in an increase in the total settling zone, but a minimum spacing is required by the critical velocity in the lamella that creates the turbulences and the risk of resuspending solids (Dudley 1994). It is recommended that the flow in the lamella follows a Reynolds number of less than 2,000 and a Froude number greater than 10^{-5} (Fischerstrom 1955). When calculating the Reynolds number (Re), the hydraulic radius (rH, lamella wet surface cross-section) must be used in the equation:

$$Re = \frac{4vrH}{\gamma}$$

where v is the water velocity, γ is the kinematic viscosity, and rH is the hydraulic radius. With this definition, laminar flow occurs at Reynolds numbers less than 500. In the context of a flow in a sedimentation tank, the Froude number (Fr) is defined as:

$$Fr = \frac{v}{\sqrt{grH}}$$

Decreasing the lamella angle increases the total settling surface.

The retention time in lamella settlers is greatly reduced to less than an hour or even less than 15 min compared to horizontal settlers. Setting up lamellae considerably reduces the Reynolds number, thus limiting turbulence, which has a direct impact on the residual turbidity.

The small size of the lamella settler systems allows them to be installed in buildings, thus facilitating the work of the operators. It also prevents the development of algae on the walls of the units and also prevents them from being less sensitive to bad weather (verglas, ice).

The Hazen velocities used for urban wastewater are between 0.8 and 1.2 $m.h^{-1}$ for the peak dry time flow.

With a conventional or lamella settler on domestic water, the total percentage of removed suspended solids is 50–65%, and that of the removed BOD_5 is 20–35%.

In primary settling, lamella systems can operate with and without reagents. The choice depends on the quality of the required settled water and the compactness of the tanks. Particle retention is therefore theoretically independent of the height of the tank.

The sludges are scraped out by means of a bottom scraper driven by chains which convey them towards one or two hoppers.

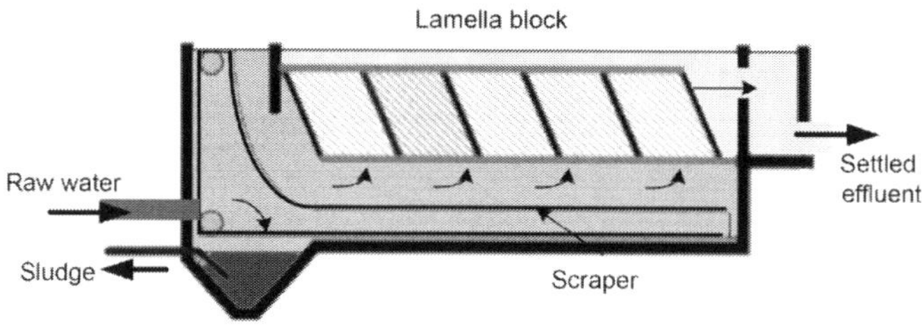

Figure 5.21. *Countercurrent lamella settler*

5.7.2.1. *Implementation of lamellar settlers*

Two categories of reagent lamella settling processes focus the market:

– classical lamella processes with mirror velocity between 8 and 15 $m.h^{-1}$;

– lamella processes with sludge contact solids or weighted flocs operating at high velocities (between 30 and 200 $m.h^{-1}$).

5.7.2.1.1. Lamella settling without addition of reagents

The suggested hydraulic loads are 10–15 $m.h^{-1}$ at peak flow. The potential for lamellae to clog is higher due to the accumulation of sticky SS; they require a reasonably uniform distribution of water resulting in a uniform distribution of flow between the lamellae. Maintenance requirements are higher for these settlers due to the need for regular cleaning. The results presented in Table 5.10 with a lamella settler (Multiflo™) concern treatment plants without chemical reagents injected into the unit.

The results obtained are as follows:

– water treatment plant (France):

Utilization rate (flow rate) = 48–57% of nominal design

– results period 1: velocity 8.7 $m.h^{-1}$, 169 mg $SS.L^{-1}$ inlet , 43% efficiency for SS_t (from 15 to 75% depending on the day) (SS_t = total SS);

– results period 2: 250–300 mg $SS.L^{-1}$ inlet, 57% efficiency for SS_t (from 30 to 80% depending on the day).

Parameters	**Units**	**Values**
Maximum flow rate	$m^3.h^{-1}$	10,980
Number of lines	–	4
Unit diameter	M	15
Unit surface at the mirror	m^2	176.6
Total water height	M	7.32
Water volume	m^3	1,292
HRT retention time	Min	28.2
Sludge pump flow rate	$m^3.h^{-1}$	75
Number of pumps per settler	–	1 + 1
Maximal mirror velocity	$m.h^{-1}$	15.5
Average mirror velocity	$m.h^{-1}$	12
Inlet SS	$mg.L^{-1}$	245
Outlet SS	$mg.L^{-1}$	102
SS reduction percentage	%	58.4
Extracted sludge concentration	$g.L^{-1}$	15

Table 5.10. *Example of dimensioning a lamella settler without reagents (Multiflo™ mono)*

The SS removal efficiency is:

– highly variable from day to day;

– not correlated with settling velocities (between 5 and 13 $m.h^{-1}$);

– higher in dry weather periods when the SS concentration in influents is higher.

An example on a lamella settler without reagents is given with the treatment plant of the Paris region (France).

The SS composition shows that 80% of the SS_t were SS_d = 80%.

At an average velocity of 12 $m.h^{-1}$, the SS_t efficiencies range between 60% and 74%. These results can be considered as very efficient with regard to the average SS_t efficiency curve of a lamella settler without reagents (see Figure 5.22).

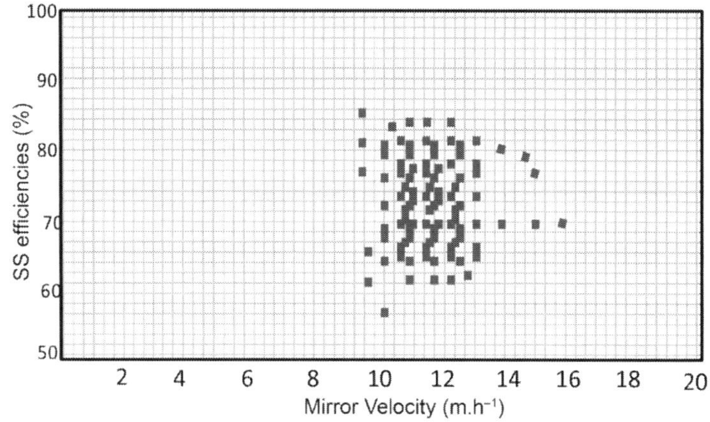

Figure 5.22. *SS efficiencies in relation to mirror velocity on a lamella settler without reagents*

5.7.2.1.2. Al Wathba treatment plant (United Arab Emirates)

Parameters	Units	Values
Dry weather daily flow	$m^3.d^{-1}$	322,100
Average hourly flow	$m^3.h^{-1}$	13,420
Rainy weather peak flow	$m^3.h^{-1}$	17,300
Number of Multiflo™	–	6
Unit length	m	16
Unit width	m	16
Usable mirror surface (unit)	m^2	170
Total mirror surface	m^2	1,020
Mirror velocity (average)	$m.h^{-1}$	13.15
Peak velocity	$m.h^{-1}$	17
Total height	m	7.6
Water height above the lamella	m	0.78
Lamella projected length	m	1.5
Settler volume	m^3	1,825

Table 5.11. *Multiflo™ dimensioning*

Considered assumptions

- Minimum raw water temperature: 27°C;
- Average raw water temperature: 30°C;
- Maximal raw water temperature: 36°C;
- N-NO3: 0 $mg.L^{-1}$;
- COD/BOD: 1.8–2.2;
- SS/COD: 0.45–0.55;
- VSS/SS: 78%;
- Settleable SS/total SS: 70–80%;
- Soluble COD/total COD: 40%;
- Non-biodegradable soluble COD: < 23 $mg.L^{-1}$;
- Non-biodegradable soluble NTK: < 1 $mg.L^{-1}$.

Raw water qualities are shown in Figure 5.23.

The average removal with Multiflo™ with no added reagents is depicted in Figure 5.24. An average SS removal of 70%, 50% with COD and 30% with BOD_5, is observed.

Figure 5.26 summarizes the average SS efficiencies obtained with a lamellar settler with no addition of reagents, as a function of the flow velocity.

This curve may change depending on the settleable SS/total SS ratio. For the Wathba plant, this ratio favorable to settling resulted in an average SS removal of 70%.

5.7.2.2. *Lamella settling with reagents*

This treatment consists of removing the stability of colloidal materials existing in the water by causing the neutralization of their surface electrical charge and the agglomeration of these particles to form a large settleable flake called a "floc".

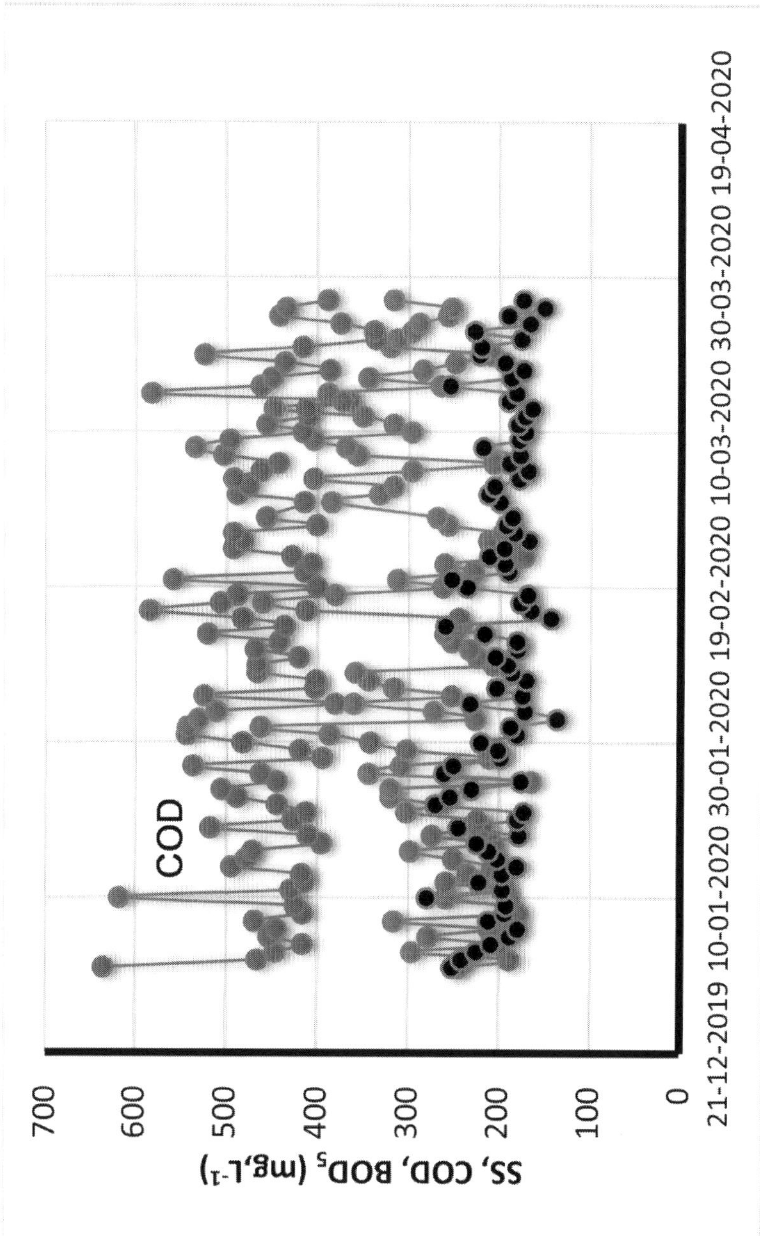

Figure 5.23. *Variations in raw water quality at the Al Wathba plant*

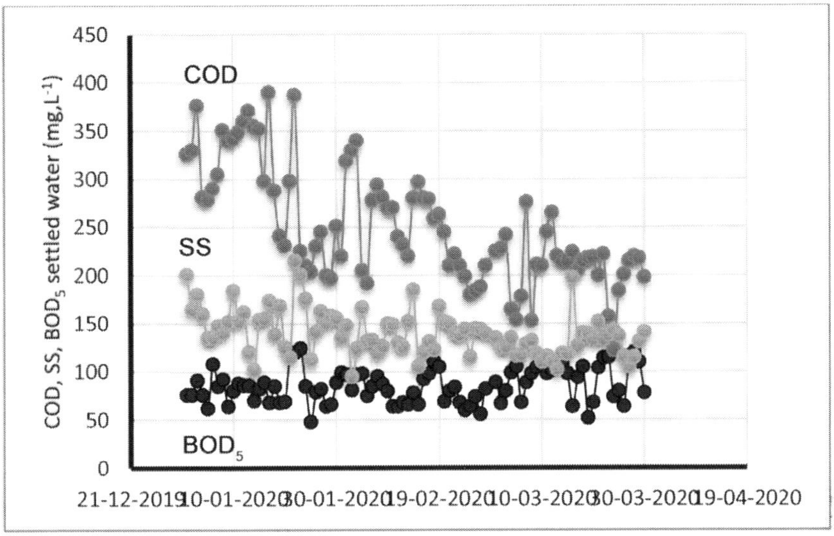

Figure 5.24. *Variations in the quality of settled water at the Al Wathba plant*

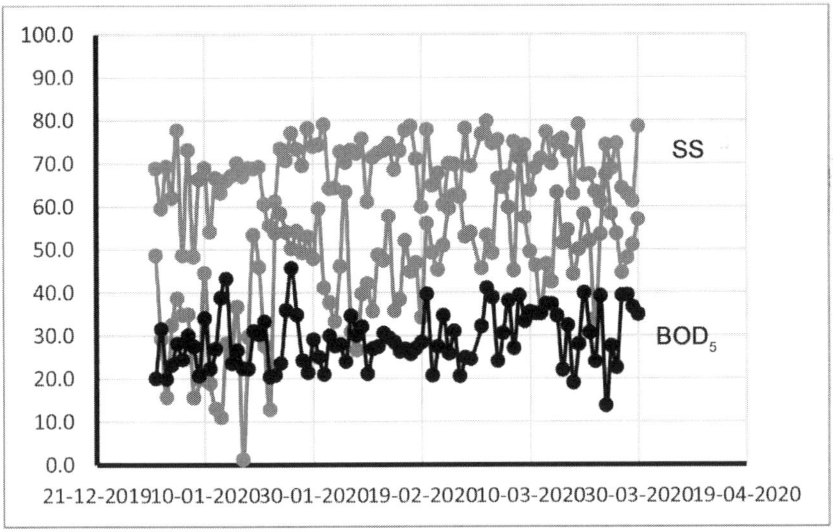

Figure 5.25. *Percentage of polluting load removal with Multiflo™*

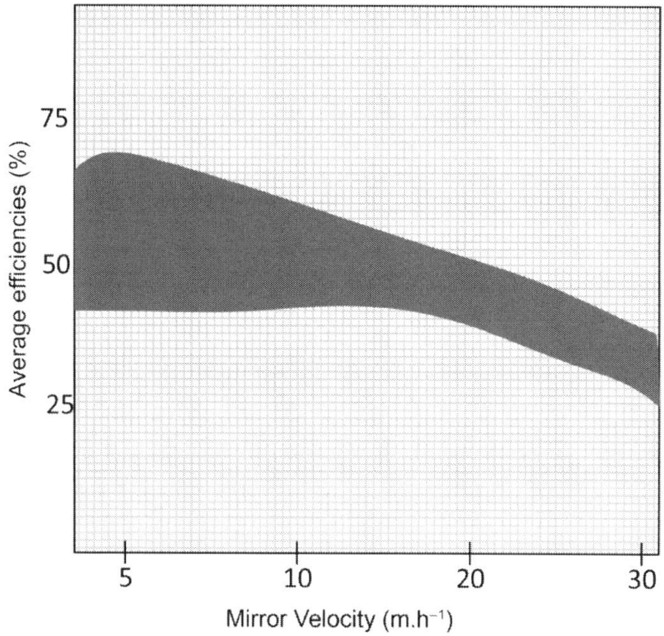

Figure 5.26. *SS average efficiency curve according to mirror velocity for a lamella settler without reagents*

The first step is coagulation and is achieved in water treatment mainly through the use of iron or aluminum salts.

The second step is flocculation and is carried out by the addition of organic polymer (long-chain macromolecule). The addition of iron or aluminum salts results in the formation of a hydroxide precipitate [$Fe(OH)_3$ or $Al(OH)_3$] and decreased alkalinity.

The reagent rates to be used are determined by flocculation tests. Adjusting the pH by adding an acid or base is sometimes necessary to obtain the optimal coagulation–flocculation pH of the reagents used: 6–7.4 for Al^{3+} and 5–8.5 for Fe^{3+}.

The choice of contact times, stirring energies and the shape of the mixing turbines is also very important and allows for substantial savings in reagent. Coagulation–flocculation, followed by settling applied to domestic water, results in removing up to more than 90% of suspended solids and 40–65% of BOD_5.

The coagulation–flocculation processes used can be conventional with lamellar settling or not, but they can also use new techniques such as sludge bed thickener settlers and ballasted floc lamellar settlers.

5.7.2.2.1. Multiflo™ settler (Veolia)

The Multiflo™ settler combines coagulation, flocculation and countercurrent lamellar settling steps in a single unit. It thus comprises in a single unit the zones where physico-chemical reactions (coagulation–flocculation) and the physical separation of the flocs (lamellar settling) take place.

The Multiflo™ settler as applied to wastewater is categorized into two types:

– the Multiflo Duo™ with Turbomix™ (optional), hoppers or harrowed scraper;

– the Multiflo Trio™ with sludge bed, sludge recirculation and Turbomix™.

In both cases, the flocculated water is then distributed under the modules through an aperture which as a particularity extends over the entire width of the device (flow baffle).

Multiflo Duo™ settler

The Multiflo Duo™ lamella settler has three main properties:

– occupies the smallest space possible using as high a settling velocity as possible;

– retains as many suspended particles as possible;

– concentrates the collected sludge.

The settling zone is equipped with hexagonal modules, consisting of inclined (60°) lamella. Therefore, two adjoining lamellae form a module composed of hexagonal cross-sectional ducts. At the bottom of the lamella, the floc settling mechanism occurs continuously despite the upflow water current. This is explained by the fact that in this area, the floc that falls is made up of large clumps that have formed by accumulation, sliding along the plates. The upflow velocity of the water does not interrupt their settling.

To prevent floc shearing during the interaction of water and floc currents, the lamellae of the Multiflo Duo™ settler are close together (< 45 mm). This small spacing improves the cut-off threshold while avoiding shearing phenomena in flocs when they slide towards the bottom of the settler.

The water flows from the bottom to top between the lamellae in the opposite direction (countercurrent) to the flocs, which are settled on the lamellae and slide downwards under the effect of gravity. The flocs arrive under the lamellae, whose total surface is calculated in such a way that with the laminar flow (Reynolds number < 200) that is established, the gravitation forces allow the flocs settled on the lamellae to slide.

The juxtaposition of inclined and aerodynamic lamellae creates independent modules, thus forming as many elementary settling units. The lamella surface also improves hydraulic distribution and reduces turbulence effects, thus facilitating the settling of less settleable flocs.

The lamellae provide a very large settling area, which corresponds to the projected surface of all the lamellae, and not only to the surface of the exposed water as for conventional settlers. This results in a space footprint 10 times lower than that of a non-lamella settling unit.

The small size of the mirror surface and the sludge accumulation area, as well as the distribution of internal flows, ensure a very homogeneous and regular operation. The feeding area of the settling part covers its entire width. In the longitudinal direction, it is included between the flooded spillway and the wall that separates the lamella zone.

The Multiflo Duo™ settler involves three trade-offs:

– an optimal lamella height (1.50 m) with the need not to resuspend the settled floc;

– the gravity flow discharge of the settled floc and the use of an optimal angle of inclination;

– a quick reaction to raw water variability in order to continuously produce high-quality settled water.

Settled water can be recovered by perforated pipes or overflow chutes with trapezoidal or rectangular notches. Dimensioning parameters are summarized in Table 5.12.

The Multiflo Duo™ settler is specially adapted for wastewater primary treatment (150–800 $mg.L^{-1}$ SS) and produces settled water of less than 40 $mg.L^{-1}$ because of the reagents.

Parameters	Multiflo Duo™	
	Polymer-free clarification	**Polymer-based clarification**
Polymer use	No	Yes
Maximum mirror velocity ($m.h^{-1}$)	8	12–15
Sludge recovery	Hoppers or harrowed circular scraper	Hoppers or harrowed circular scraper
Sludge SS concentration ($g.L^{-1}$)	10–20 $g.L^{-1}$	10–20 $g.L^{-1}$
Flocculation time (according to T °C)	10–30 min	10–30 min
Spacing between lamellae	45 mm	45 mm
Lamella length	1 m	1 m
TPS/Sm lamella	10 m^2/m^2	10 m^2/m^2
Expected performance (raw water → settled water)	500 → 150 $mg.L^{-1}$	500 → 40 $mg.L^{-1}$

Table 5.12. *Basic parameters of the Multiflo Duo™ settler*

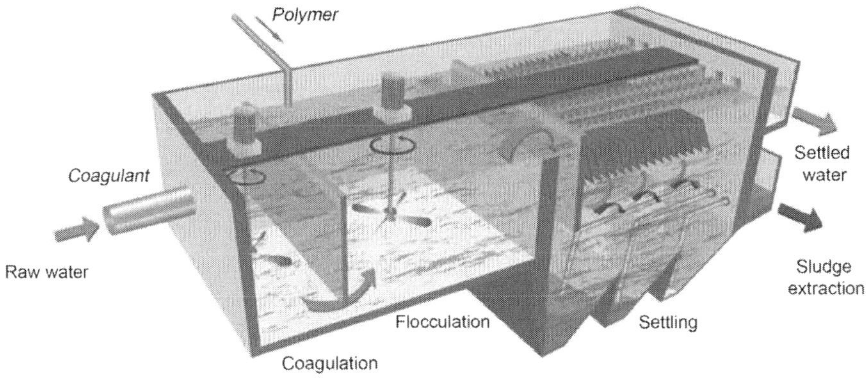

Figure 5.27. *Multiflo Duo™ hopper settler*

Example of the design of a Multiflo Duo™

– The quality of raw wastewater is presented in Table 5.13.

	Raw water quality		
	Average	95 percentile	Maximum
COD ($mg.L^{-1}$)	600	1,150	1,290
BOD_5 ($mg.L^{-1}$)	305	520	680
SS ($mg.L^{-1}$)	274	400	600

Table 5.13. *Raw water quality*

– The dimensioning is shown in Table 5.14.

Parameters	Unit	Average flow rate	Peak flow
Hourly flow rate	$m^3.h^{-1}$	8.374	20.624
Number of Multiflo Duo™ settlers	–	4	
Coagulation tanks	–	4	
Unit coagulation tank volume	m^3	137	
Volume total	m^3	548	
Average contact time	min	3.9	1.6
Coagulant	–	$Al_2(SO_4)_3$	
Flocculation tanks	–	4	
Unit flocculation tank volume	m^3	940	
Volume total	m^3	3,760	
Average contact time	min	26.9	10.9
Flocculant	–	Anionic polymer	
Settlers	–	4	
Settling unit mirror surface	m^2	231	
Total mirror surface	m^2	924	
Mirror velocity in N	$m.h^{-1}$	9.1	22.3
Hazen velocity	$m.h^{-1}$	0.87	2.14
Mirror velocity in N^{-1}	$m.h^{-1}$	12.1	29.8
Height	m	6.75	
Lamella angle of inclination	°	60	
Total projected surface (TPS) per settler	m^2	2,407	
TPS/mirror surface	–	10.42	
Lamella length	m	1.44	

Table 5.14. *Dimensioning of a Multiflo Duo™*

– For raw water with a BOD_5 concentration between 200 and 400 $mg.L^{-1}$ and an SS concentration between 200 and 350 $mg.L^{-1}$, a dosage of 3.33 $mg.L^{-1}$ of Al is required (equivalent to 20 $mg.L^{-1}$of $FeCL_3$ pure product).

– For raw water with a BOD concentration > 400 $mg.L^{-1}$ and an SS concentration > 350 $mg.L^{-1}$, a dosage of 6.65 $mg.L^{-1}$ of Al is required (equivalent to 40 $mg.L^{-1}$ of $FeCL_3$ pure product).

– The Multiflo Duo™ is equipped with an SS measurement in the inlet channel in order to allow an injection of reagent to be adjusted to the inlet concentration.

– Liquid alum sulfate coagulant (6% Al_2O_3) is prepared from a powder product with 15.6% Al_2O_3. The dosing pumps are dimensioned for variable Al dosages between 0.8 and 6.9 $mg.L^{-1}$.

– The anionic polymer is dosed on average at 0.5 $mg.L^{-1}$. The dosing pumps are dimensioned for variable dosages between 0.3 and 1 $mg.L^{-1}$.

5.7.2.2.2. Ballasted settling

Ballasted settling is a physico-chemical treatment that uses a mineral support that is continuously recycled and associated with chemical reagents to improve the settling performance of suspended solids based on a binding between flocs and the mineral support. The fast formation of heavy flocs ensures a very short particle sedimentation time and allows clarification to perform up to 10 times faster than conventional clarification; it thus makes it possible to treat water at a much higher rate than that achieved by traditional unit processes.

Ballasted settling units work by adding a coagulant (iron salts). Combined with added chemicals, microsand is very effective in ensuring binding with flocs by reducing flocculation and settling time (Liao et al. 1999). For example, the Actiflo® process works with hydraulic velocities of 50–200 $m.h^{-1}$ while allowing for removal of suspended solids in the order of 80–95% (Tarallo et al. 1998).

The Multiflo Trio™ settler (Veolia)

The Multiflo Trio™ settler is a thickener-settler with the same basic characteristics as the Multiflo Duo™ settler, namely a coagulation zone, a flocculation zone and a settling zone. The water flows from bottom to top between the lamella in the opposite direction (countercurrent) to the flocs, which are settled on the plates and slide downwards under the effect of gravity.

The flocs flow underneath the lamellae, the total area of which is calculated in order for the laminar flow (Reynolds number < 200) that is established to be such that the gravitational forces allow the flocs settled on the lamellar plates to slide.

The Multiflo Trio™ settler is characterized by the presence of a fluidized sludge at the flocculator level that accelerates flocculation as it goes through the raw water, with the production of denser and more easily separable flocs. By definition, the flocculation step is intended to form large aggregates enabling them to quickly settle.

The recirculation of sludge at the flocculation tank level results in an increased number of particles and accelerates collisions. It improves the effectiveness of the treatment.

Sludge recirculation (5–10%) leads to an increased concentration in the flocculation tank and thus an increase of the final concentration for the same sludge retention time. This results in densified flocs that increase the sedimentation rate. Recirculation avoids, in periods of low turbidity, overdosing the reagent that operators tend to introduce in order to increase the floc volume and ensure better settling.

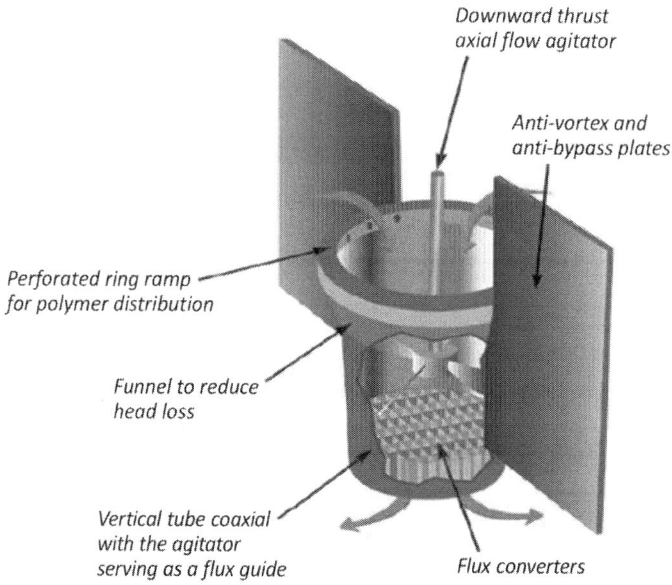

Figure 5.28. *Turbomix™ diagram*

The efficiency of the Multiflo Trio™ settler is also based on optimized flocculation that includes specific equipment such as the Turbomix™ mixer. The flocculation reaction is then very effective when all the coagulant and polymer react to destabilize the colloid particles. Turbomix™ (see Figure 5.28) thus allows homogeneity to be introduced in the flocculation tank by ensuring polymer distribution, avoiding bypasses and dead zones in this tank, and makes it possible to increase the pumping rate at equal power.

Figure 5.29. *Turbomix™ schematic and installation in a plant in Shanghai-Bailonggang (China)*

Figure 5.30. *Mirror surface*

Another characteristic of the Multiflo Trio™ settler is the presence of an extended settling zone located between the feeding zone and the sludge collection zone. Prolonged sedimentation leading to a significant reduction in the quantity of sludge admitted between lamellae makes it possible to increase the upflow velocity (and therefore the flow velocity between the lamellae) without any risk of resuspension of the layer of flocs settled on the lamellae. In addition, the large opening between the extended settling zone and the sludge recovery zone results in a reduction in the velocity under the lamellae as well as in any risk of resuspension of the deposited sludge.

The settling zone is equipped with hexagonal modules, consisting of inclined (60°) lamella. Between the lamella and the level of the sludge blanket d, there is a 1–2 m zone which provides transient flow to stimulate the separation between water and sludge.

The Multiflo Trio™ settler is characterized by the formation of a sludge blanket. The flow in the sludge blanket is uniformly divided in the settler due to the coalescence that occurs during this step and which enables collisions. The flow of sludge arriving within the sludge blanket does not act like a plug flow piston because the low velocity conditions of the mixture (scraper) reduce the intensity of turbulent fluctuations and create a homogeneity that tends to reduce the porosity of the sludge blanket due to the agglomeration of the sludge. This homogeneity results in establishing a balance between all the forces present (pressure forces, hydraulic forces), which endows this sludge layer with an overall cohesion.

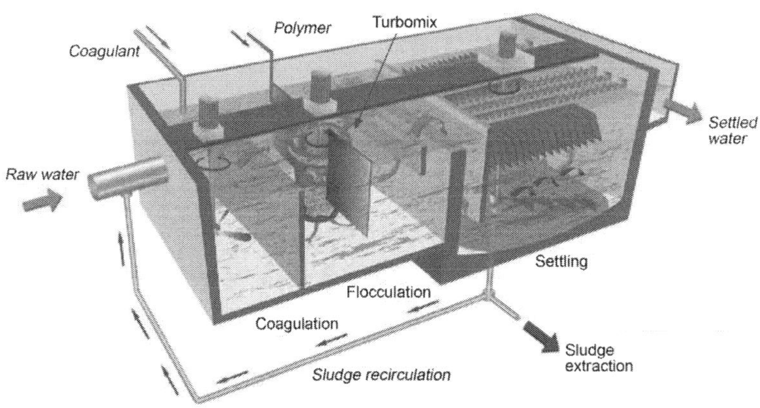

Figure 5.31. *Multiflo Trio™ settler*

In addition, moderate turbulence fluctuations continue to favor flocculation by approximating the flocs to one another at a minimal distance under the action of van der Waals forces. One of the interesting consequences is that the sludge blanket behaves like a filter in which the new flocculated particles originating from the flocculation tank are retained on the surface of the flakes already present in the sludge blanket. The flow velocity transiting the sludge contact solids is calculated so that it does not disturb either the agglomeration of the flocs or the density of the flocs in the sludge blanket.

Parameter	Unit	Value
Raw water temperature	°C	10.0
Average raw water flow	$m^3.h^{-1}$	3,002
Multiflo coagulation tank	–	1
Coagulation feed flow rate	$m^3.h^{-1}$	3,002
Coagulation reaction time	Min	3.75
Multiflo flocculation tank	–	1
Flocculation feed flow rate	$m^3.h^{-1}$	3,002
Flocculation reaction time	Min	11.2
Number of operating Multiflo units	–	2
Multiflo feed flow rate	$m^3.h^{-1}$	1,501
Internal diameter	M	13
Mirror surface	m^2	114
Mirror velocity	$m.h^{-1}$	13.16
Hazen velocity	$m.h^{-1}$	0.88
Inclined plate length	M	1.04
Orthogonal spacing	Mm	40
Inclination angle	°	60
Total projected surface (TPS)	m^2	1,710
TPS/Sm	–	15
Water height above the lamella	M	0.78
Total water height	M	6.23
Settler volume	m^3	994
Retention time in Multiflo	Min	39.7

Table 5.15. *Multiflo Trio™ settler dimensioning parameters*

The deepening of the zone under the lamella is intended to increase the retention time of the sludge, to capture a significant part of the SS not retained in the first zone and to avoid sludge resuspension by the inflowing current.

As such, the Multiflo Trio™ settler includes equipment that allows for:

– harrowing the sludge compartment in order to eliminate water pockets and ensure a mixture between the flocs already settled and those arriving from the flocculation tank;

– a syncopated extraction of thickened sludge (rate-duration). The extraction times are dependent on the mass of the produced sludge, thus avoiding the extraction of liquid sludge. Due to the permanent supply of flocculated colloids and reagents introduced, the level of the sludge bed gradually rises if purges are not carried out regularly to ensure a constant height of sludge. Sludge purges are performed from the bottom of the settler without generating dynamic disturbances of the sedimentation.

Example of a structure in operation: Shanghai-Bailonggang (China) case study

Parameter	Wet weather peak	Dry weather average	Dry weather peak
Wastewater flow rate ($m^3.h^{-1}$)	78,660	50,000	65,016
Number of settlers	12	12	12
Flow rate per settler ($m^3.h^{-1}$)	6,555	4,167	5,418
Mirror unit (m^2)	187	187	187
Mirror velocity ($m.h^{-1}$)	35.02	22.26	28.94

Table 5.16. *Design of Multiflo Trio™ settlers (Veolia)*

	COD	BOD_5	SS	$N\text{-}NH_4$	$P\text{-}PO_4$
Settler inlet ($mg.L^{-1}$)	320	130	170	30	5
Settled water ($mg.L^{-1}$)	180	70	28	30	1
Removal percentage	44	46	83.5	0	80

Table 5.17. *Multiflo Trio™ performance*

Sludge recirculating settler–thickener: the Densadeg™ (Degrémont)

This settling process combines chemical coagulation, sludge recirculation, lamella settling, sludge thickening and recycling.

The process includes four functions (see Figure 5.32):

– Coagulation: the raw water enters a flash mix zone where a coagulant is added, causing agglomeration of the colloidal particles.

– Flocculation: in a second zone, the coagulated water is brought into contact with a flocculating agent and the thickened, recirculated sludge coming from the settling/thickening zone. The recirculated sludge accelerates the flocculation process and insures a dense and homogeneous floc. The transition to the settling stage is done via a piston reactor with an upward current in which the flocculation continues.

– Clarification/settling: in this zone, due to their size and particularly their density, the flocs settle at the bottom of the settling tank. The clarified water is separated from the descending sludge and goes back up through the lamellar modules. The sloping lamellae, which are configured in a honeycomb pattern, act like a refining stage, trapping the lighter less dense solids, which have not settled. The uniform speeds at which they go through prevent any sludge going back up: the clarified water has the same quality over the entire recovery area.

– Thickening and densification of sludge: the settled sludge is progressively thickened at the bottom of the settling tank using a floor scraper and picket fence. Thanks to the density of the flocs, the thickening is very rapid and superior to that obtained from other solid contact settlers. A part of the thickened sludge is recycled towards the coagulation zone, and the surplus is periodically evacuated (Memento Degremont 2004).

This unit therefore combines the recirculating sludge flocculation principle with lamellar settling while integrating a thickener. It consists (see Figure 5.32) of a flash (or rapid) mixer (1), an agitated flocculator (2) followed by a piston flocculator including a sludge blanket (3), a settling tank (4) in which 90% of the flocs settle and thicken, a lamella zone (5) topped by crenellated chutes enabling a uniform collection of the treated water and a circuit (6) for returning the thickened sludge to the flocculator. The coagulated wastewater enters the flocculation tank where a polymer is injected into the mixture. At the end of this stage, the wastewater enters the settling zone where the floating materials (oils and greases) are collected from the surface. The inclined lamellae facilitate flocs sliding, which settle at the bottom of the unit and ensure solid–liquid separation. The settled sludge is thickened, and part of this sludge is recycled towards the flocculation tank.

The operation principle of this settler–thickener is as follows: concentrated sludge is recycled at a low flow rate towards the flocculation tank. The recirculated sludge multiplies the contact surfaces and accelerates the phenomena of floc agglomeration. Flocculation in the Densadeg™ is optimal if the spatial occupancy of the flocculates inside the agitated flocculator is about 10%.

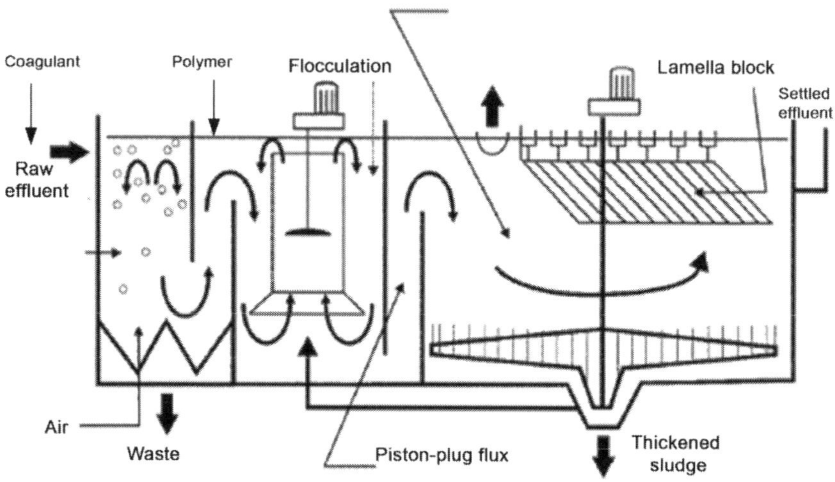

Figure 5.32. *Densadeg™ operating diagram*

On the settling side, three functions predominate:

– a hindered settling of most of the sludge that relies on the size of the flocs and their density, thus enabling very high settling velocities;

– an additional refining of the settling process in the lamella modules that trap the residual flocs;

– a thickening with a bottom scraper and a harrow, allowing very quick thickening that is much more effective than that obtained with other older sludge contact settlers.

The polymer can be injected either into the sludge recirculation line or under the blades of the mixer turbine. The recirculated sludge and high solid concentration optimize the flocculation process while the polymer facilitates the densification of the coagulated particles.

The addition of an organic flocculant (polymer) between the flocculation reactor and sludge recirculation reduces the injected concentration and enables the production of thick sludge. The design procedures for rapid mixing with sludge blanket settlers are similar to those used for conventional sedimentation with G values of 200–300 sec^{-1} and hydraulic retention times of 30–60 seconds. The flocculation times are significantly shorter than those of the conventional process (12 minutes). They are only about six minutes, and there is an intensity of the mixture in the flocculation zone that keeps the floc in suspension without breaking the newly formed flocs. Part of the settled sludge (2–6% of the flow) is recycled back into the flocculation area. By increasing the number of particles in the water, sludge recirculation increases the flocculation rate and the increase in floc density intensifies the sedimentation rates.

The depth of the settling zone is significantly lower than that of traditional settlers. The stabilization time is shorter compared to that of traditional clarifiers, and allows for much more significant crossing flows. This zone also acts as a thickener and allows increasing the concentration of solids. Part of the sludge is sent to the flocculation tank to be used as a "ballast" material.

The solids extracted from the structure generally consist of between 3 and 6% dry matter. Additional thickening is not required in most cases.

Densadeg™ applications include:

1) improved primary settling;

2) improved secondary clarification;

3) reducing peak flow for stormwater.

The advantages of this process are:

– reduced settler and clarifier areas;

– a wide range of flows without reducing efficiency;

– a clear improvement in settling and the reduction of solids compared to traditional systems.

The Densadeg™ is also used as a primary treatment plant for rehabilitation or replacement projects.

The hydraulic velocities announced can reach up to 100 $m.h^{-1}$ under peak flow conditions. When a Densadeg™ starts operating outside the rainy period, its full efficiency is reached after 20–30 minutes of operation. In rainy periods, its full effectiveness is almost immediately reached.

In dry weather, hydraulic loads vary between 7 $m.h^{-1}$ for average flows and 30 $m.h^{-1}$ for peak flows.

Figure 5.33. *Densadeg™ (South of France)*

BOD_5 and SS removal falls into the 70%–80% range depending on the inlet conditions.

The installation of a suction tube mixer enables integral mixing in the flocculation zone. Chemical coagulation combined with sludge recycling forms dense, heavy flocs that quickly settle in the clarifier/thickener. The thickener produces the concentrated sludge which is returned to the flocculation tank turbine. Sludge concentrations range between 40 and 60 $g.L^{-1}$. The sludge is intermittently extracted.

	Primary wastewater	**Tertiary wastewater**	**Rainwater**
Maximal velocity ($m.h^{-1}$)	35	30	90
Extraction sludge concentration ($g.L^{-1}$)	30–100	25–50	30–100
Outlet SS concentration ($mg.L^{-1}$)	< 30	< 10	< 50

Table 5.18. *Densadeg™ dimensioning and performance*

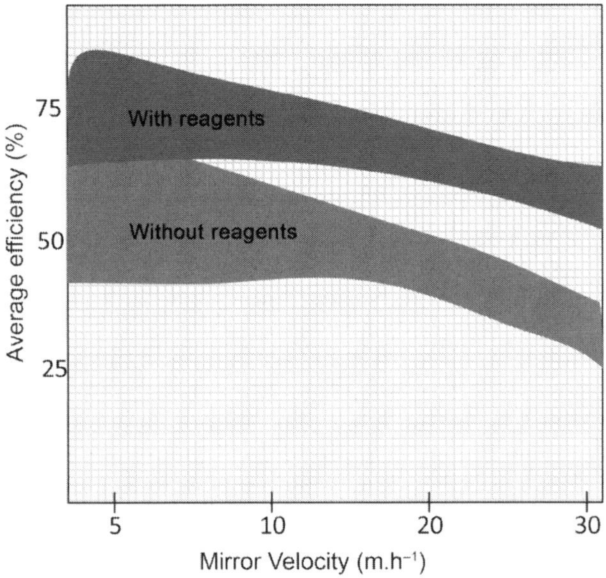

Figure 5.34. *Settler performance with and without reagents versus flow velocity*

Figure 5.34 shows that the addition of a coagulant (Fe^{3+} or Al^{3+}) and a flocculant (anionic polymer) greatly improves settler performance and water discharges containing low SS proportions. This is significant when settled water is then directed to biofilters (see Volume 2, Chapter 3).

Examples of results obtained with lamellar sludge bed settlers

– Cergy treatment plant (France), Multiflo™:

- mirror velocity: 11.8 m.h^{-1} (design: 24.3 m.h^{-1});

- pure $FeCl_3$ doses: 18.6 mg.L^{-1} (12.3–27.1 mg.L); proposed design with 25 mg.L;

- SS_t removal: 79% (design: 77.7);

- factory inlet SS_t: 539 mg.L^{-1};

- outlet SS_t: 26–80 mg.L^{-1};

- average efficiencies obtained: 85–95%;

- inlet COD: 920 mg.L^{-1};

- settled water COD: 300–400 mg.L^{-1};

- average efficiencies obtained: 52–67%;

- Fe dosage: 15 → 20 mg.L^{-1};

- no influence of sludge recirculation on performance;

- optimal flocculation: 1–1.2 mg.l^{-1} anionic polymer;

- degraded flocculation at low polymer dose (≤ 0.5 mg.L^{-1});

- thickened sludge concentration: 60 g.L^{-1}.

– Colombes treatment plant (France): Densadeg™. The Densadeg™ settler is installed with biofilters (Biostyr) with 34 m.h^{-1} at peak and 22 m.h^{-1} with mini HRT:

- operates at an average of 21 m.h^{-1} (26 m.h^{-1} at peak) and 35 m.h^{-1} HRT (30 m.h^{-1} at peak) with excellent results;

- stable performance over the tested range (17–26 m.h^{-1} on average per month);

- inlet total SS: 150–205 mg.L^{-1};

- outlet total SS: 20–30 mg.L^{-1};

- total SS efficiency: 82–92% and 88% on average;

- pure iron dosage: 10–18 mg.L^{-1} (or 29–52 mg.L^{-1} pur $FeCl_3$, that is, 35–40 mg.L^{-1} average);

- concentration of thickened sludge (discontinuous extraction): on average 57 g.L^{-1}.

Ballasted floc settler (Actiflo®, Veolia)

Ballasted floc is the result of using materials introduced during the flocculation phase, which are intended to increase the size and the weight of the flocs created by coagulation. This results in faster settling. The materials used as ballast can be mineral, inert or active. Several materials are used: microsand, various kinds of clays, magnetite, activated carbon, etc., as well as the sludges (flocs) extracted from the settling which are recycled.

Mineral materials induce higher settling rates than recirculated flocs, due to their higher density. This combination increases the settling velocities by a factor varying between 10 and 50 depending on the type of settler (lamellar or not). The units are more compact, and the concentration of suspended solids in the settled water is smaller.

Ballasted flocculation is therefore a physico-chemical clarification process operating at high speed, and involving the fixation of flocs or suspended solids on a ballast with the help of a polymer.

During the flocculation step, the Smoluchowski expression shows that any increase in the particles tends to reduce the contact time:

$$Ln\frac{N}{No} = -\frac{4}{3}\alpha Vp\, Gt$$

with:

– N and No: number of free colloidal particles at times t and to;

– α: effective collision frequency factor;

– Vp: volume of particles per suspension volume;

– G: velocity gradient;

– t: contact time.

Good quality flocculation is obtained when the Ln N/No ratio decreases. This explains why, for a low concentration of suspended solids (SS), that floc recirculation or particle injection can increase the probability of aggregation. In addition, the increase in floc diameter favors flocculation and increases the chances of floc mixing. The particularity of microsand is that it provides a large number of particles (contact area), improving the flocculation rate; in addition, it acts as ballast that accelerates floc settling.

Therefore, suspended solids present in the wastewater, which is destabilized by addition of coagulant, binds to microsand particles or recycled flocs via polymer bonds. These agglomerations of large particles turn into high-density flocs. Enough mixing energy and retention time allow agglomerates to trap random flocs and produce water with a low SS concentration.

Figure 5.35. *Ballasted flocculation with microsand enabling fast settling*

The Actiflo® process consists of three steps:

– a rapid coagulation zone into which the coagulant is injected;

– a rapid mixing tank, called an injection tank, which receives the recycled sand and the polymer;

– a third tank, called the maturation tank, in which the flocs grow due to the effect of metal hydroxides and sand.

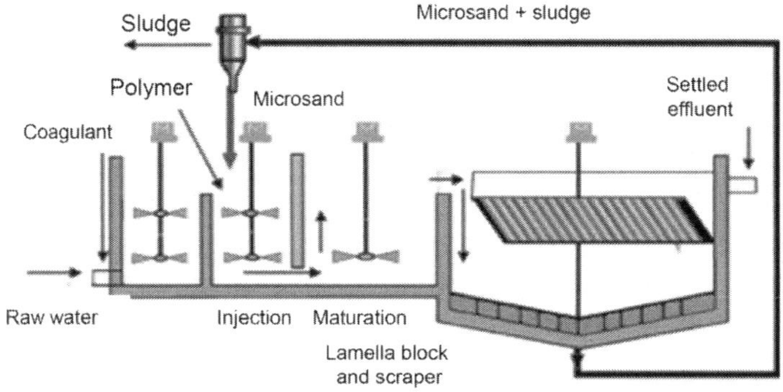

Figure 5.36. *Schematic operation diagram of the Actiflo® process*

The raw water enters a coagulation tank into which the coagulant is injected, which allows destabilizing the colloidal materials and trapping inert matter. The coagulant can also be directly injected upstream of the tank (in-line injection). The contact time is 2 min although it can reach as high as 3–4 min at very low temperatures. The velocity gradient lies between 400 and 700 s^{-1}.

The coagulated water then flows through an injection tank into which the recirculated microsand and polymer are injected. Rapid mixing ensures the dispersion of microsand and polymer, making it possible for microflocs and other suspended solids to attach the microsand. The particle size of the microsand is 135 μm (effective size). The doses of sand vary between 3 and 5 $kg.m^{-3}$. The reactivity of its large specific surface area initiates agglomeration reactions with the solids present (microflocs and suspended solids) and the polymer. The contact time is usually 2 min.

The microsand mass travels through the flocculation zone, dragging the newly created aggregates, through bonds with the polymer, metal hydroxides and suspended solids/colloidal materials (see Figure 5.37). The ballast provided by the microsand injection acts as a starter and initiator for flocculation. The coagulated materials clump around each grain of sand. Since it has a very highly developed surface area due to its size (T.E.135 μm), it offers immediate responsiveness even at cold temperatures.

The microsand content to be ensured in the flocculation tank ranges between 5 and 10 g.L^{-1}.

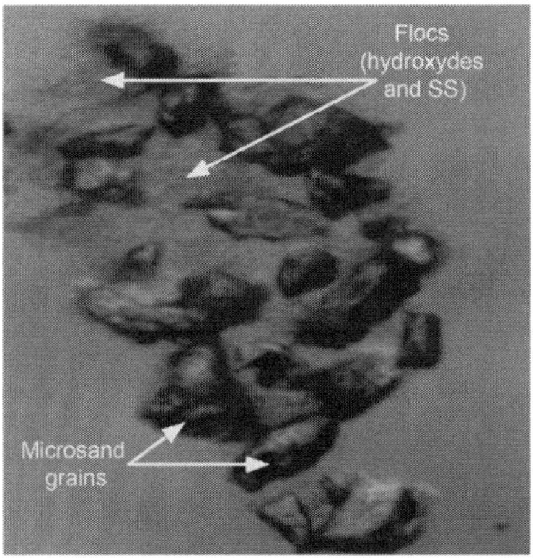

Figure 5.37. *Microsand–floc aggregate (suspended hydroxides–materials/colloidal materials)*

Sand increases the number and size of particles in the water, which has a positive effect on flocculation kinetics. The mixture then goes through a third tank, called the maturation tank, where floc maturation takes place. The retention time is 6 min. The velocity gradient G lies between 120 and 165 s^{-1}. Finally, the liquid/solid separation occurs in the countercurrent lamellar settler.

The settled water is discharged by a collector towards the biological basins. The settled flocs are collected either by a scraper to a central sump or at the bottom of a hopper.

In the presence of microsand, which forms the ballast, the resulting sludge is collected at the bottom of the settler to be pumped into hydrocyclones, where the sludge is separated from the microsand by centrifugation. The hydrocyclone creates a centrifugal vortex that separates the microsand from the sludge. In this way, almost all of the microsand is recovered. The sulfur containing the sand is recycled in the flocculation tank. The overflows, loaded with sludge, are discharged to the sludge treatment.

Specific equipment is used to control the doses of reagents (coagulant and flocculant) as well as the microsand concentration in the recirculation stream, the recycling rate and the velocity gradient in the coagulation and flocculation tanks. The presence of microsand in high concentrations makes it possible to:

– increase the probability of particle collisions;

– increase the exchange area and therefore the adsorption capacity compared to conventional flocculation;

– form solid and dense ballasted flocs that withstand strong agitation and then quickly settle.

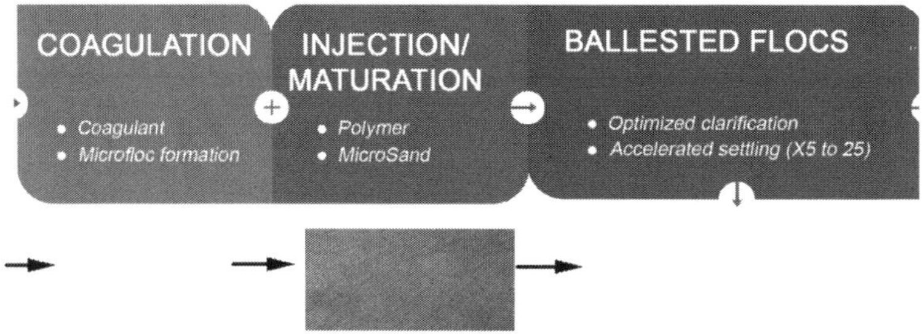

Figure 5.38. *Principle of ballasted flocculation in Actiflo®*

Due to this mechanism and the physico-chemical interactions that take place in the successive tanks, flocculation and settling require very short retention times that lead to a very compact process.

This means that the stirring time and the volume concentration of solids in the flocculation zone are much longer than in a conventional flocculator. The flocculation rate is therefore theoretically increased despite a small velocity gradient. This makes the flocs larger and the Hazen velocity in the settling zone increases.

The reaction times in the three tanks are:

– coagulation time = maturation time/2;

– injection time = maturation time/2;

– maturation time (min) = $3 \times 80/v_{mirror}$ ($m.h^{-1}$).

Ballasted floc settlers with an inert material such as microsand require an overall reaction time of less than eight minutes.

The most important advantage of the settler is its retention time, which is less than 15 minutes at the designed nominal rate. This time is significantly shorter than in a conventional process. The use of ballasted flocculation therefore requires much less space. The space required occupies about 5% of the space used by a conventional settling system. This technology makes it possible to produce water with qualities equal to or better than water produced by means of conventional processes but using a much more confined space.

The mechanisms of ballasted flocculation with microsand can be explained with the relation:

$$\frac{dN}{dt} = -KaGN + Kb\,G^2$$

where N is the particle concentration, Ka is the aggregation constant, Kb is the deterioration constant (s), G is the velocity gradient (s^{-1}) and t is the contact time (s).

The first term of the equation is a measure of the magnitude of aggregation, while the second term is a measure of the magnitude of aggregate deterioration due to turbulence, too violent collisions and temperature. The results show that with Actiflo®, the optimal velocity gradient is about 3–10 times higher (150–500 s^{-1}) than in a conventional process (50 s^{-1}). The deterioration constant Kb is 25 times smaller than for conventional treatment. This relation shows that the aggregation forces are significantly higher than the destruction and shear forces of the flocs.

The results obtained show that the flocs formed are more robust and more cohesive. They effectively withstand the increase in the velocity gradient that is necessary due to the mass of microsand in the reaction tank.

The flow is laminar and unidirectional. The floc is subject to only two forces: the feed velocity (upward velocity) and gravitation. There is a very strong interaction between the particles, and there is no influence due to floc shape (or other particles present in the water) because gravitation is conditioned and managed by the mass of microsand present in the mixture.

The concentration of microsand ($kg.m^{-3}$) is calculated by the equation:

$$m = \frac{1.5 \text{ x } Qr}{Qi}$$

The density of the microsand in the mixture is equal to 1.5 $kg.m^{-3}$, the recirculation rate Qr ($m^3.hr^{-1}$) is variable between 5 and 8%. Qi is the raw water inlet flow rate in Actiflo® ($m^3.h^{-1}$).

Settling rate of ballasted flocs

The Stokes equation shows that for a diameter of agglomerated flocs of 300 μm, the settling velocity of the flocs is 81 $m.h^{-1}$. This means that Actiflo® can operate in wastewater at velocities of at least 60 $m.h^{-1}$:

$$vs = \frac{g\, d^2(\rho s - \rho water)}{18\,\mu}$$

where g is the gravity constant (9.81 $m^{2}.s^{-1}$), μ is the dynamic viscosity (1.13 10^{-3} $kg.m^{-1}.s^{-1}$), d is the floc diameter (m), ρs is the density of the agglomerated flocs ($kg.m^{-3}$) and ρwater is the density of the water ($kg.m^{-3}$).

For an aggregate (floc–sand) of 200 μm, the theoretical settling rate deduced from the Stokes relation is:

$$vs = \frac{g(\rho s - \rho water)dp^2}{18\mu}$$

$$vs = \frac{9.81\ m.s^{-2}(2{,}500 - 1{,}000)(200\ 10^{-6})^2}{18 \text{ x } 1.3\ 10^{-3}\ )}$$

$$= 0.0251\ m.s^{-1} = 90.55\ m.h^{-1}$$

where ρs is the sand density at 2,500 $kg.m^{-3}$, ρwater is the water density at 1,000 $kg.m^{-3}$, the diameter of the ballasted flocs is 200 μm (200 10^{-6} m) and μ is the dynamic viscosity of 1.3 10^{-3} $kg.m^{-1}.s^{-1}$.

Reynolds number

$$Re = \frac{\Phi v p d p}{\mu} = \frac{(2.5)(1{,}000)(0.0348)\ (200\ 10^{-6})}{1.3\ 10^{-3}} = 13.3$$

Φ is the form factor: 2.5.

Since Re > 1, Newton's law must be used to access the settling rate in the transition zone.

The drag coefficient C_D is given by:

$$CD = \frac{24}{Re} + \frac{3}{\sqrt{Re}} + 0.34$$

That is:

$$\text{CD} = \frac{24}{13.3} + \frac{3}{\sqrt{13.3}} + 0.34 = 2.968$$

We then have:

$$\text{vs} = \sqrt{\frac{4g(\rho s - \rho water)d}{3\ \text{CD}\Phi} = \sqrt{\frac{4(9.81)(2.5-1)(200\ 10^{-6})}{3\ (2.968)(2.5)}}} = 0.0236\ \text{m.s}^{-1} = 85.1\ \text{m.h}^{-1}$$

The velocities obtained in both cases are quite similar. The variation in shape coefficient (5–10%) can affect the accuracy of the calculation. The separation efficiency of microsand is normally greater than 99.9%. The overall sand loss with Actiflo® also includes microsand leaks into the settled water, which depend on the quality of the flocculation and the settling velocity. It is sometimes compensated by adding a non-negligible amount of microsand to the raw water and by the addition of new sand. The choice of hydrocyclones essentially depends on the flow rate to be cyclonically separated and the grain size of the sand used.

The choice of coagulant depends on the characteristics of the water to be treated, the goal of the treatment (SS removal, advanced BOD removal) and pH constraints. The doses to be applied are to be defined and optimized by Jar tests.

The amount of flocs generated by the doses of reagent depends on a K-value which is relative to the hydroxides bound to the coagulant (see section 5.7.2).

The loss of sand with sludge (in hydrocyclone overflow) varies according to the effective size and dose of sand, the recirculation rate, the operating conditions of the hydrocyclones and the sludge concentration to be discharged.

The ballasted flocculation process has significant advantages, including:

– small floor area;

– fast start-up times to reach maximum efficiency in less than 20 minutes;

– the wide range of streams that can be processed with high efficiencies of SS and BOD removal;

– well-clarified water in terms of suspended solids.

Ballasted flocculation is currently used worldwide in coagulation–flocculation–settling plants.

The Actiflo® ballasted floc process makes possible the treatment of peak flows in rainy weather and integrates well into treatment systems for applications where rainy weather requires immediate intervention. It is often applied as primary settling, in secondary clarification and in tertiary treatment.

It occupies a small space compared to conventional processes and has a very short start-up time and an immediate response in terms of efficiency.

Not affected by the fluctuations in the quality of raw wastewater, it enables at the same time improved removal of mineral and organic matter in particulate form. The advantage of such a procedure resides in the fact that for large plants, in periods of dry weather, since the circulation velocity is small, the number of units in operation can be limited. The remaining units can be used in tertiary treatment for the refining of suspended solids or phosphorus.

In summary, the main interest of Actiflo® is its low response time. It allows:

– sudden variations in the load and quality of the raw water (from 0 to more than 1,000 $mg.L^{-1}$ of SS);

– significant variations in the flow rate to be treated (from 30 to 100%);

– a very fast start-up, even at the nominal flow rate;

– treatment of different types of water in the same unit (alternating or mixing).

The rapid mix design procedures for ballasted flocculation are similar to those used for conventional sedimentation with G-values from 200 to 300 s^{-1} and hydraulic retention times of 30–60 seconds. The flocculation times for ballasted flocculation are significantly shorter than those of conventional processes.

Hydraulic retention times for ballasted flocculation fall within the range of 1–3 minutes, resulting in Gt values (G being the mean velocity gradient ($L.s^{-1}$) and t being the hydraulic retention time (s)) between 6,000 and 20,000.

In Actiflo®, hydraulic velocities are estimated between 60 and 80 $m.h^{-1}$ at average flow rates and can reach 100–150 $m.h^{-1}$ at peak flow. Tests at 180 $m.h^{-1}$ showed that the system was still working satisfactorily.

Actiflo® can address flow rates comprised between 10 and 100% of its nominal capacity, making it possible to respond to very high instantaneous rates.

Application	Mirror velocity ($m.h^{-1}$)	Comment
Primary before biofilter	80	Nominal dimensioning
Primary before biofilter + biofilter wash water	80	
Primary before activated mud	120	
Rain	120	
	150	Degraded mode

Table 5.19. *Applicable mirror velocities according to the treatment*

Coagulation dosages in Actiflo® when applied to wastewater generally range between 40 and 60 $mg.L^{-1}$ with ferric chloride, 80–85 $mg.L^{-1}$ with ferric sulfate, 60–70 $mg.L^{-1}$ with alum, and 45–100 $mg.L^{-1}$ with polyaluminum chloride. Anionic polymer dosages vary between 0.8 and 1.2 $mg.L^{-1}$.

The flocculant is normally injected in two points:

– first, 25–50% in the hydrocyclone underflow, ensuring that the mixture with the sand is homogeneous;

– the remainder into the inlet aperture in the maturation tank, using a perforated injection ramp placed at 2/3 of the height (the injection ramp must cover the width of the inlet aperture).

The flow distribution between the two points must be adjusted by manual valves and rotameters, even at low flow rates.

Type of treatment	Minimum dosing for correct flocculation		Dosing for conventional SS discharge	
	Coagulant ($mg.L^{-1}$ pure $FeCl_3$)	Polymer ($mg.L^{-1}$)	Coagulant ($mg.L^{-1}$ pure $FeCl_3$)	Polymer ($mg.L^{-1}$)
Primary	20	1	40–60	1–1.2
Primary + biofilter wash water	20	1	40–60	1–1.2
Rain	30	1	≥ 60	1–1.2

Table 5.20. *Reagent dosing in Actiflo®*

The sludge concentrations originating from Actiflo® vary between 3 and 8 g.L^{-1}, depending on the SS concentration at the inlet and the recirculation rate.

	Primary	**Rain**	**Tertiary (after biofilter or clarifier)**	
	Reduction	**Reduction**	**Reduction**	**Expected effluent**
SS	85% (65–90)	85% (80–95)	70% (50–80)	10–20 mg.L^{-1}
Total COD	60% (50–80)	60% (50–80)	50% (40–60)	< 50 mg.L^{-1}.
Soluble COD	30% (25–50)	25% (20–30)	25% (20–30)	–
Total BOD_5	–	70% (60–80)	–	< 15 mg.L^{-1}
Soluble BOD_5	–	20% (15–25)	–	–
NTK	15% (10–30)	25% (20–35)	–	–
Ptot	According to the coagulant dose			
P-PO_4	According to the coagulant dose			
Heavy metals	–	50–100% according to metal	–	–
Fat	50% (HEM)	–	–	–
E. coli	–	1.5–2 log (95-99%)	–	–
Helminth eggs	< 1 u/liter for 50% of the time	–	< 1 u/liter for 100% of the time	

Table 5.21. *Actiflo® performance with respect to wastewater parameters*

Polymers, in particular, often need to be aged or activated before use and can then only remain active for a limited period of time. Because wet weather events cannot be predicted with precision, polymer feeding solutions must be regularly prepared and replaced.

There is an upper limit to the amount of sand that can be amalgamated to the floc, and any amount of sand above this dose contributes little to additional efficacy. Sand concentrations in the maturation tank (flocculation) vary from 2 to 12 g.L^{-1} with a supplementary dose of about 3–5 mg.L^{-1} for wastewater treatment.

Table 5.21 summarizes the Actiflo® performance with regard to mineral and organic substances present in wastewater.

Primary settling	Effectiveness (%)		Peak velocity ($m.h^{-1}$)	Reference
	BOD_5	SS		
Conventional reagent-free	25–30	40–50	3.4–5.0	WEF (1998)
Conventional chemically assisted	45–65	60–85	3.0–5.0	Odegaard (1998)
Lamellar Multiflo Trio™ type	40–60	60–90	30–40	Rogalla et al. (1992)
Densadeg™	40–60	75–90	25–100	Metcalf and Eddy, Inc. (2003)
Actiflo® ballasted flocs	40–60	75–90	100–130	Young and Edwards (2000)

Table 5.22. *Comparison of performances and flow velocities between the different kinds of processes used for primary settling*

Ballasted flocculation processes (microsand or sludge bed) require greater expertise compared to the operation of conventional systems because the addition of ballast requires close monitoring of the material recirculation conditions.

The short retention times require a quick intervention by the operator in order to maintain the required operating conditions to obtain the final quality of the treated water (coagulant, polymer, sludge extraction, recirculation rate).

For wet weather applications, the maintenance of the units is more demanding than that of conventional systems that only store excess water volumes and then treat them, outside the rainy period.

Since Densadeg™ requires the formation of a sludge bed dense enough to trap solid matters, a latency time is needed before optimal SS removal rates can be achieved. As a result, the latency time is strongly dependent on the quantity of solids entering the unit.

Parameter	Actiflo®	Densadeg™ and Multiflo Trio™
Ballast product	Microsand 135 μm	Sludge
Hydraulic load ($m.h^{-1}$)	> 130	> 30
Overall retention time	< 20	< 40
Concentration of extracted sludge ($g.L^{-1}$)	< 5	40-80
Recirculation rate	Hydrocyclone 5–8%	Pump 10–12%
Startup	Instantaneous	< 30 min
Chemical reagents	Coagulant + flocculant	Coagulant + flocculant

Table 5.23. *Ballasted floc settler design comparison*

Parameters	Densadeg™	Actiflo®	Densadeg™	Actiflo®
Daily flow	226,000 $m^3.d^{-1}$		342,000 $m^3.d^{-1}$	
Number of units	2	2	3	2
Unit capacity ($m^3.d^{-1}$)	113,000	113,000	114,000	171,000
Width (m)	11.6	7.6	11.6	8.8
Length (m)	22	18	22	20
Depth (m)	8.5–10.4	6	8.5–10.4	7.6
Number of pumps	7	6	10	6
Pumping capacity ($m^3.h^{-1}$)	142	150	142	230

Table 5.24. *Densadeg™ and Actiflo® dimensioning elements*

Sludge bed and ballasted floc processes offer significantly reduced footprints and often increased pollutant removal with remarkable efficiency gains.

For all settlers using chemical reagents, Table 5.25 summarizes the characteristics of the main coagulants.

Parameter	Conventional	Ballasted floc system
Removed SS rate	50–60%	80–95%
Removed COD rate	25–40%	65–85%
Removed BOD_5 rate	25–40%	50–75%
Removed Pt rate	< 25%	90–95%
Removed heavy metals rate	–	60–90%
Chemical reagents	–	Coagulant, flocculant, ballast

Table 5.25. *Performance of conventional settlers and ballasted floc settlers*

Coagulant	Characteristics
Ferric chloride ($FeCl_3$) Pure product atomic mass = 162.5	$FeCl_3$ pure product content = 41% (w/w) Solution density = 1,430 kg.m^{-3}
Ferrous sulfate ($FeSO_4.7H_2O$) Anhydrous product atomic mass = 152 Hydrated product atomic mass = 278	Hydrated product concentration in solution = 250 g.L^{-1}
Liquid aluminum sulfate ($Al_2(SO_4)_3$) Anhydrous product atomic mass = 342 Alumina atomic mass = 102	Al_2O_3 content = 8.2% (w/w) Technical product (liquid) density = 1,330 kg.m^{-3}
Solid aluminum sulfate ($Al_2(SO_4)_3$)	Al_2O_3 content in the technical product (solid) = 17.5% (w/w) Technical product maximal solubility (at 20°C) = 650 g.l^{-1}

Table 5.26. *Chemical reagents used in lamellar settlers*

EXAMPLES.–

– Ferric chloride (liquid):

- dose = 50 g.m^{-3} pure $FeCl_3$ = 17 g.m^{-3} Fe (50*56/162.5) = 122 g.m^{-3} 41% technical product (= 50/0.41);

- volume dose = 122 g.m^{-3}/1,430 g.L^{-1} = 0.085 l/m^3.

– Ferrous sulfate (solid):

- dose = 50 g.m^{-3} pure $FeSO_4$ = 18.4 g.m^{-3} Fe (50*56/152) = 91.4 g.m^{-3} solid or technical product (= 50*278/152);

- volume dose = 91.4 g.m^{-3}/250 g.L^{-1} = 0.366 l.m^{-3}.

– Aluminum sulfate (liquid):

- dose = 50 $g.m^{-3}$ of Al_2O_3 alumina = 26.5 $g.m^{-3}$ Al (50*54/102) = 610 $g.m^{-3}$ 8.2% liquid product (= 50/0.082);

- volume dose = 610 $g.m^{-3}$/1,330 $g.L^{-1}$ = 0.459 $l.m^{-3}$.

– Aluminum sulfate (solid):

- dose = 50 $g.m^{-3}$ of Al_2O_3 alumina = 26.5 $g.m^{-3}$ Al (50*54/102) = 286 $g.m^{-3}$ 17.5% solid product (= 50/0.175);

- volume dose = 286 $g.m^{-3}$/$g.L^{-1}$ = 0.440 $l.m^{-3}$.

5.8. References

Aiguier, E., Chebbo, G., Bertrand-Krajewski, J.-L., Hedges, P., Tyack, N. (1996). Methods for determining the settling velocity profiles of solids in storm sewage. *Water Sci. Technol.*, 33(9), 117–125.

Aiguier, E., Chebbo, G., Bertrand-Krajewski, J.-L., Gagne, B., Hedges, P. (1998). Analysis of the methods for determining the settling characteristics of sewage and stormwater solids. *Water Sci. Technol.*, 37(1), 53–60.

Albertson, O.E. and Walz, T. (1997). Optimizing primary clarification and thickening. *Water Environ. Technol.*, 9(12), 41.

Andersen, K.K. (1996). Secondary sedimentation of activated sludge by a lamella separator placed in an oxidation ditch. *Eur. Water Pollut. Control*, 6, 37–44.

Andoh, R. (1998). Improving environmental quality using hydrodynamic vortex separators. *Water Qual. Int.*, 47–51.

Argaman, Y. and Kaufman, W.F. (1970). Turbulence and flocculation. *J. Sanit. Eng. Div., Proc. Am. Soc. Civ. Eng.*, 96, 223.

Béchac, J.P., Boutin, P., Mercier, B., Nuer, P. (1983). *Traitement des eaux usées*. Eyrolles, Paris.

Boeglin, J.-C. (2002). Traitements physico-chimiques de la pollution insoluble. *Techniques de l'Ingénieur*, G1270.

Buer, T. (2002). Enhancement of activated sludge plants by lamella in aeration tanks and secondary clarifiers. In *International Water Association World Water Congress*. International Water Association, London.

Buer, T., Herbst, H., Marggraff, M. (2000). Enhancement of activated sludge plants by lamella in aeration tanks and secondary clarifiers. In *Aeration Conference on Applied Techniques to Optimise Nutrient Removal and Aeration Efficiency*, Helsinki, 2–7 October.

Camp, T.R. (1945). Sedimentation and the design of settling tanks. *Trans. Am. Soc. Civ. Eng.*, 111, 895–958.

Chang, F.Y., Gutshall, M., Skradski, R. (1998). Microsand enhanced flocculation for wastewater treatment: Results from pilot plant studies in primary, tertiary, and CSO applications. In *Proceedings of the 71st Annual Water Environment Federation Technical Exposition and Conference*. Water Environment Federation, Alexandria, 825–832.

Clesceri, L.S., Greenberg, A.E., Eaton, A.D. (1998). *Standard Methods for the Examination of Water and Wastewater*, 20th edition. American Public Health Association, American Water Works Association, Water Environment Federation, Washington, D.C.

Degremont (2004). *Memento technique de l'eau. Tomes 1 et 2*. Degremont, Paris.

Desbos, G., Laplace, C., Rogalla, F. (1990). Extended coagulation for reagent and space savings with wastewater lamella settling. In *Chemical Water and Wastewater Treatment: Proceedings of the 4th Gothenburg Symposium*, Hahn, H.H. and Klute, R. (eds). Springer Verlag, Berlin, 139–150.

Droste, R.L. (1997). *Theory and Practice of Water and Wastewater Treatment*. John Wiley & Sons, New York.

Dudley, J., Mulhall, E., Solomon, S. (1994). Primary treatment processes, to reduce BOD and suspended solids – Final report, UM 1458. WRc, Swindon.

Fischerstrom, C.N.H. (1955). Sedimentation in rectangular basins. *Sanit. Eng. Div. Am. Soc. Civ. Eng.*, 81, 1–29.

Gaid, A. (2002). Traitement des eaux résiduaires. *Techniques de l'Ingénieur*, C5 220-v2.

Guibelin, E., Delsalle, F., Binot, P. (1994). High speed settling of stormwater with microsand. In *Chemical Water and Wastewater Treatment III: Proceedings of the 6th Gothenburg Symposium*, Hahn, H.H. and Klute, R. (eds). Springer Verlag, Berlin, 112–132.

Harleman, D. and Murcott, S. (2001). An innovative approach to urban wastewater treatment in the developing world. *Water*, 21, 44–48.

Hazen, A. (1904). On sedimentation. *Trans. Am. Soc. Civ. Eng.*, 53, 63.

Hazen, A., Pearsons, G.W., Weston, R.S., Fuller, G.W. (1904). Discussion on sedimentation. *J. Assoc. Eng. Soc.*, 53, 45–88.

Jolis, D. and Ahmad, M.-L. (2001). Pilot testing of high rate clarification technology in San Francisco. In *Proceedings of the 74th Annual Water Environment Federation Technical Exposition and Conference*. Water Environment Federation, Alexandria.

Keller, J., Schultze, M., Wagner, D. (2001). Lawrence Kansas: Detailed design. Issues for a ballasted flocculation system. In *Proceedings of the 74th Annual Water Environment Federation Technical Exposition and Conference*. Water Environment Federation, Alexandria.

Larsson, H.F. (ed.) (1986). Solid/liquid separation equipment scale-up. In *Lamella Separators*. Uplands Press, Croydon, 215–218.

Liao, S.L., Ding, Y., Fran, C.-Y., Field, R., Chan, P.C., Dresnack, R. (1999). High rate microcarrier-weighted coagulation for treating weather flow. In *Water Environment Technology Symposium*, New Orleans.

Metcalf & Eddy, Inc. (1991). *Wastewater Engineering: Treatment, Disposal, and Reuse.* McGraw-Hill, New York.

Metcalf & Eddy, Inc. (2003). *Wastewater Engineering: Treatment and Reuse*. McGraw-Hill, New York.

Metcalf & Eddy, Inc. (2014). *Wastewater Engineering: Treatment and Reuse*. McGraw-Hill, New York.

Odegaard, H. (1998). Optimized particle separation in the primary step of wastewater treatment. *Water Sci. Technol.*, 37, 43–53.

Parker, D.S., Esquer, M., Hetherington, M., Malik, A., Robison, D., Wahlberg, E.J., Wang, J.K. (2000). Assessment and optimization of a chemically enhanced primary treatment system. In *Proceedings of the 73rd Annual Water Environment Federation Technical Exposition and Conference*. Water Environment Federation, Alexandria.

Parker, D.S., Barnard, J., Daigger, G.T., Tekippe, R.J., Wahlberg, E.J. (2001). The future of chemically enhanced primary treatment: Evolution not revolution. *Water*, 21, 49–56.

Rich, L.G. (1961). *Unit Operations of Sanitary Engineering*. John Wiley & Sons, New York.

Rogalla, F., Roudon, G., Sibony, J., Blondeau, F. (1992). Trade-off between physico-chemical lamella separators and aerated biofiltration. In *Chemical Water and Wastewater Treatment II: Proceedings of the 5th Gothenburg Symposium 1992*, Klute, R. and Hahn, H.H. (eds). Springer Verlag, Berlin, 429–445.

Scruggs, C. and Wallis-Lage, C. (2001). Ballasted flocculation: A wet weather treatment solution? In *Proceedings of the 74th Annual Water Environment Federation Technical Exposition and Conference*. Water Environment Federation, Alexandria.

Sibony, J. (1981). Clarification with microsand seeding a state of the art. *Water Res.*, 15, 1281–1290.

Tarallo, S., Bowen, M.W., Riddick, A.J., Sathiamoorthy, S. (1998). High rate treatment of CSO/SSO flows using a high density solids contact clarifier thickener. Results from a pilot study. In *Water Environment and Technology Symposium*, New Orleans.

Tebbutt, T.H.Y. and Christoulas, D.G. (1975). Performance relationship for primary sedimentation. *Water Res.*, 9, 347.

Wahlberg, E.J., Wang, J.K., Merrill, M.S., Morris, J.L., Kido, W.H., Swanson, R.S., Finger, O., Phillips, D.A. (1997). Primary sedimentation: It's performing better than you think. In *Proceedings of the 70th Annual Water Environment Federation Technical Exposition and Conference*. Water Environment Federation, Alexandria.

Wahlberg, E.J., Wunder, D.B., Fuchs, D.C., Voigt, C.M. (1999). Chemically assisted primary treatment: A new approach to evaluating enhanced suspended solids removal. In *Proceedings 72nd Annual Water Environment Federation Technical Exposition and Conference*. Water Environment Federation, Alexandria.

Water Environment Federation (1998). *Design of Municipal Wastewater Treatment Plants*, 4th edition. Manual of practice no. 8. Water Environment Federation, Alexandria.

Water Environment Federation (WEF) (2005). *Clarifier Design*, 2nd edition. Manual of practice no. FD-8. McGraw-Hill, New York.

Young, J.C. and Edwards, F.G. (2000). Fundamentals of ballasted flocculation reactions. In *Proceedings 73rd Annual Water Environment Federation Technical Exposition and Conference*. Water Environment Federation, Alexandria.

6

Biological Treatment with Activated Sludge

6.1. Background information on microbial organisms

The discharging of wastewater into a receiving environment (river, lake, stream, etc.) incurs various reactions that are likely to transform the ecological balance of the environment.

In principle, the purpose of a biological wastewater treatment system is to replicate the decomposition processes of pollutants that occur naturally in the receiving environment. The treatment system is a confined area that provides favorable conditions for the quick degradation of these contaminants.

The treatment is done in different stages. In the first stage, soluble or colloidal contaminants are consumed by microorganisms, which transform them into cellular constituents. Secondly, these microorganisms and other materials in excess can be mechanically separated from the purified water. The activated sludge system is therefore a large culture of suspended microorganisms that consume pollutants to transform them into cell mass and partially into CO_2. In order to accelerate the degradation process of organic contaminants, the biomass is concentrated by the solid–liquid separation process and then recycled back into the system, resulting in a differentiation between the hydraulic retention time (τ) and the retention time of the sludge (sludge age: G). This endows the treatment with good stability and high efficiency.

Solid–liquid separation is achieved through various settling or filtration processes. The reduction of certain nutrients, such as phosphorus during biological phosphate removal or nitrogen during nitrification and denitrification, is carried out by exposing biomass to different toxic cycles. A wide variety of activated sludge system configurations have been developed in recent decades in order to meet various treatment goals. The description of all these variants is too elaborate to be included in this book, but the reader may refer to the reference works listed at the end of this chapter if necessary.

Chemical transformations play a secondary role compared to biological transformations. In fact, they are most often limited to acidity or free alkalinity fixation reactions, or even to the oxidation of strongly reducing substances (sulfides, for example) and of some products resulting from the anaerobic decomposition of organic impurities.

Biological transformations are more important inasmuch as they involve a multitude of reactions of various kinds.

The reaction rates depend on various physicochemical factors (temperature, pressure, dissolved oxygen, nature of the organic products). Biologic reactions involve both organic substances and mineral substances such as phosphates and nitrates, which are assimilated for the synthesis of new cells.

Among the various organisms responsible for biological phenomena, bacteria are the most significant and represent the largest number. A bacterium can be thought of as a system that uses a substrate to produce energy, create new cells and release metabolic products.

These different operations are common to all bacteria, allowing them to transform substrate into gaseous, liquid or solid products.

The operation of cell synthesis is called anabolism, and that relating to the production of energy and formation of degradation products is called catabolism. These two operations together form the metabolism.

The specific conditions of the environment may condition the nature of the reaction: it will be aerobic in the presence of oxygen and anaerobic in the absence of oxygen. As such, there are aerobic bacteria and anaerobic bacteria. It should be noted, however, that the degradation rate of organic matter is higher in aerobic environments, which is why biological purification plants generally operate in the presence of oxygen.

Bacteria can also be classified into two other categories:

– autotrophic bacteria, which, under the action of solar energy, use carbon dioxide and salts dissolved in water to synthesize new cells; for example:

$$106CO_2 + 90\ H_2O + 16\ NO_3^- + PO_4^{3-} \rightarrow C_{106}\ H_{180}\ O_{45}\ N_{16}\ P + 77\ O_2$$

– heterotrophic bacteria that use organic substances for cell synthesis.

The composition of wastewater shows a significant concentration of organic matter. This is the case for urban wastewater. This proportion may decrease when an industrial effluent is connected to the plant (fertilizer manufacturing plants, for example). Furthermore, the nature of the bacteria responsible for the treatment of urban wastewater is heterotrophic.

Biological degradation is achieved in two almost simultaneous phases:

– a very fast absorption phase, during which the organic substances are absorbed on the outer membrane of the cells;

– a slower oxidation phase, during which the oxidation of organic matter into decomposition products such as CO_2 and H_2O occurs.

The degradation rate depends on several parameters such as the amount of required oxygen, the total mass of microorganisms, temperature and primarily the nature of the substances to be treated. Many substances are actually very quickly degraded, while others are more slowly degraded.

For example, alcohols, phenols, sugars, amino acids and organic acids properly degrade and olefins, some detergents, ketones and aldehydes degrade less well, except after the microorganisms have adapted to these substances. Other products degrade very little or not at all, such as chlorinated hydrocarbons and humic acids. However, it seems that for the latter, an association with molecules of the same nature but more easily biodegradable, such as phenols, leads to interesting removal results.

6.2. Operating principle

The most effective method for the reduction of organic matter in wastewater is therefore its biological degradation. This is carried out through aerobic bacteria that cause direct oxidation from dissolved oxygen in the water. Or even from anaerobic bacteria where oxidation is indirect, that is, characterized by a loss of hydrogen fixed by an acceptor other than molecular oxygen.

The activated sludge process is a continuously operating system in which microorganisms are brought into contact with wastewater containing organic matter. Oxygen is injected into the mixture, providing the bacteria with the vital element for their respiratory needs.

In fact, the activated sludge system can be considered an artificial extension of natural purification phenomena. In a river or stream, the phenomena involved are identical to those present in activated sludge systems; only the concentration of microorganisms in the environment varies and, in particular, the rate of the degradation reaction.

Activated sludge systems are most often designed to be continuously operated with a biological reactor, a secondary settler and sludge return equipment for the settler and the reactor. A typical process flow diagram is shown in Figure 6.1. These systems can also be designed to be operated in sequential mode, with some time dedicated to directly settling in the reactor.

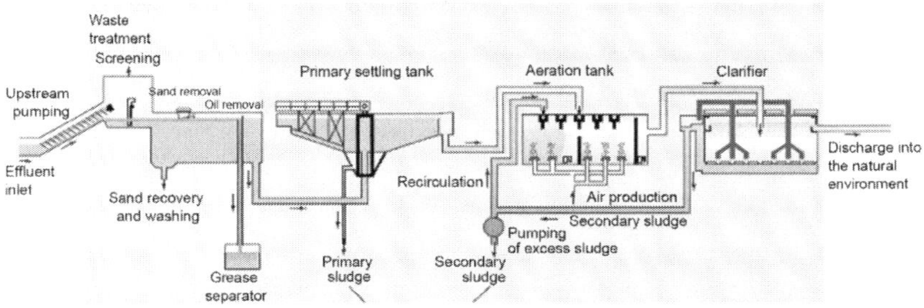

Figure 6.1. *Wastewater treatment system using activated sludge*

6.3. Theoretical bases of aerobic treatment

In the presence of abundant feed, microorganisms will quickly develop and tend to form more or less compact flaky masses. The bacterial floc thus defined will make the oxidation of biodegradable materials possible. In fact, the phenomenon is more complex, since part of the organic matter is assimilated and transformed into cellular material, while another portion is degraded by oxidation, where the recovered energy is used for cell synthesis. When food is abundant, there is then a storage phenomenon displayed by the cells, which use it according to their needs (Figure 6.2).

6.3.1. *Organic matter (biological oxygen demand) evolution in the presence of bacterial mass*

When organic matter is fed to a microbial medium, and if the effluent contains nutrients, there is a gradual evolution of the microbial mass in four main phases (Figure 6.2).

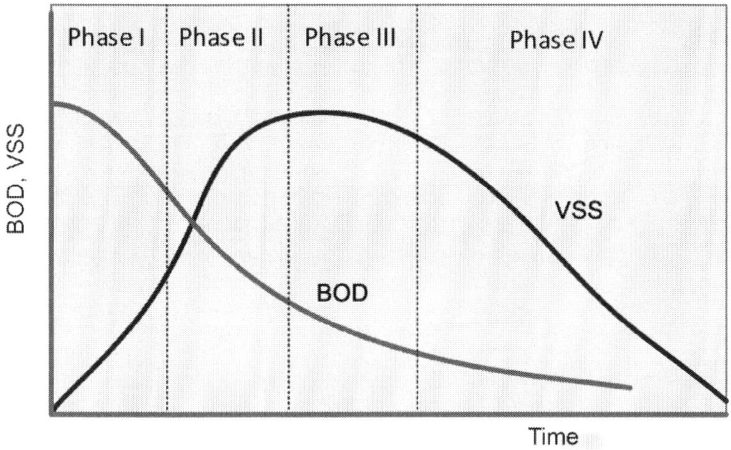

Figure 6.2. *BOD and VSS evolution in a biological tank*

During phase I, the medium rich in nutrients enables fast bacterial growth. Biological oxygen demand (BOD) quickly decreases; the VSS mass (volatile suspended solids) increases while the mass of oxygen present in the medium decreases. This is the phase of cell synthesis during which oxidized organic matter is transformed into final degradation products such as CO_2, H_2O, NO_2, etc.

During phase II, lack of nutrient causes a slowdown in bacterial growth. It is then observed that the VSS mass starts to level off. This is the so-called slow growth phase.

Phase III corresponds to a steady phase. It means that at this time, the bacteria use previously accumulated reserves.

Finally, during phase IV, the medium is poor in organic matter, which results in the death of many microorganisms. This is the endogenous phase. The oxygen supplied is then used by the bacteria for their own transformation into final products. The general scheme includes:

– fast absorption and flocculation of colloidal suspended solids and soluble organic materials by activated sludge;

– progressive oxidation and synthesis of adsorbed organic matter as well as that extracted from the solution (continuous system);

– additional aeration that leads to oxidation and dispersion of sludge particles.

Three essential points should be remembered:

– The oxidation rate of organic pollutants in wastewater is higher as the overall mass of microorganisms in contact is high.

– When the organic matter content decreases, the microorganisms carry on their activity at the expense of their own matter.

– Microorganisms having undergone an autoxidation phase return to their activity with some difficulty, which becomes more problematic if the medium is deficient in oxygen.

6.3.2. *Analytical interpretation*

The design of an activated sludge system must necessarily be based on a mass balance that takes into account the loads and contaminant fractions of the influent, the effluent, the extracted sludge and the biological kinetics in the reactor. The volume of aeration tanks must be sufficient to contain all the biomass required to maintain the substrate/microorganism ratio (F/M) that was chosen by the manufacturer. Once the mass of solids to be maintained in the reactor has been defined, the concentration of the mixed liquor is determined according to the solid–liquid separation system that has been chosen.

The contaminant degradation process in the activated sludge process is described by a first-order kinetic model based on biomass growth (Grady 1999; Metcalf and Eddy, Inc. 2003). According to this model, the substrate is consumed during biomass growth.

The mathematical interpretation of the VSS growth and BOD removal curves has been the focus of various studies. Very early on, Monod (1941) showed that "in the complex media used, the growth limit is imposed by the disappearance of a single nutrient, or at least of the group of substances that form the carbonaceous source". He defines a notion of efficiency such as:

Efficiency = total growth/medium concentration

He also adds that the relation between total growth and concentration of the substrate is linear, which makes it possible to write:

$$am = \Delta X/\Delta L$$

where am is the biomass yield.

This relation leads to a clarification of the notion of growth. Examination of the previous curve relating to the evolution of the biomass concentration in the presence of a substrate in an aerated medium prompts us to express this law in an exponential form such as:

$$dX/dt = \mu X$$

where X represents the biomass concentration present in the medium at time t, and μ is the growth rate of the bacteria.

This expression can also be expressed in exponential form:

$$X = Xo\ e^{\mu t}$$

6.3.2.1. *Study of the logarithmic growth phase*

During this phase, the substrate is abundant, and the cell development is only limited by their reproductive time. Since a fraction of the pollution is converted into new cells, it can be assumed that after a reaction time t equivalent to:

$$\Delta t = t - to$$

the biomass concentration increases from X_0 to X_1 such that:

$$X_1 = X_0 + \Delta X$$

The previously defined efficiency had made it possible to write:

$$\Delta X = am\ \Delta L, \text{ that is, } \Delta X = am\ (Lo - L1)$$

with Lo ($mg.L^{-1}$), BOD_5, as input, or still:

$$\Delta X = am\ Le$$

where Le ($mg.L^{-1}$) is the removed BOD_5 at time Δt for a biomass growth equal to ΔX.

This relation becomes:

$$am\ (Lo - Lf) = Xo\ (e^{\mu t} - 1)$$

$$\mu t = \text{Log}\ [(\frac{Lo - Lf}{Xo})\ am + 1]$$

$$\mu t = \text{Log}\ [1 + \left(\frac{amLe}{Xo}\right)]$$

The line obtained by carrying over Log $[1 + \left(\frac{amLe}{Xo}\right)]$ with respect to t gives μ.

μ is then defined as the specific growth rate. This term is often interpreted as the time necessary to double the bacterial population. It is typically used when designing the aeration tank.

6.3.2.1.1. Example of determining the aeration tank

Laboratory tests discontinuously conducted yielded the results presented in Table 6.1.

Aeration time (h)	**BOD_5 ($mg.L^{-1}$)**
0	5,400
0.5	5,054
2	4,492
4	4,050
8	2,937
16	677
24	475

Table 6.1. *BOD_5 reduction with respect to time in a biological tank*

L	**Lo – L**	$\frac{amLe}{Xo}$	**Log** $[1 + \left(\frac{amLe}{Xo}\right)]$
5,000	346	0.121	0.114
4,492	908	0.317	0.275
4,050	1,350	0.472	0.387
2,937	2,463	0.862	0.621
677	4,723	1.653	0.975
475	4,925	1.723	1.002

Table 6.2. *Calculations of the different factors of the equation of μ*

Sludge concentration in the tank is Xo = 2,000 $mg.L^{-1}$, and the biomass yield is am = 0.70.

The curve giving Log $[1 + \left(\frac{amLe}{Xo}\right))$ according to t makes it possible to derive the value of μ Xo. We get μ= 0.052.

The values given by the literature are from 0.05 to 0.13. When one wants to know the volume of the aeration tank using the value of μ, the following calculation method can be used:

– flow rate: 1,000 $m^3.d^{-1}$;

– Y, removal efficiency: 90%;

– recycled sludge concentration: 8,000 $mg.l^{-1}$.

The recycling rate is then:

$$r = \frac{X}{Xr - X} = \frac{2{,}000}{8{,}000 - 2{,}000} = 0.33$$

Taking into account the recycled BOD_5, the BOD_5 of the mixture becomes equal to:

$$L'o = \frac{Lo + r(Lo - YLo)}{1 + r}$$

$$L'o = \frac{5{,}400 + 0.33(5{,}400 - 0.9 \text{ x } 5{,}400)}{1 + 0.33} = 4{,}194 \text{ mg. L}^{-1}$$

The BOD_5 to be removed becomes: Le = 4,194 - 540 = 3,654 $mg.L^{-1}$.

Integrating these values into the equation of μ yields:

$$t = \frac{1}{\mu} \text{Log } [1 + \left(\frac{amLe}{Xo}\right)$$

$$t = \frac{1}{0.052} \text{Log } [1 + \left(\frac{0.7 \text{ x } 3{,}654}{2{,}000}\right)]$$

t = 15.8 hours

Volume of the biological tank, V = 1,000 x 15.8/24 = 658 m^3

The value of μ therefore makes it possible to determine the dimensions of the aeration tank. It can, however, only be obtained in a laboratory. An estimate is made by assuming, for a unit continuously operating, that the variation in bacterial mass does not differ by more than 10 to 15% from its average value Xa, for a short enough time t.

The value of Log (1 + am Le/Xo) can be restricted to the first term of its series expansion. Let:

$$\mu = \text{am Le/Xa t}$$

The notation μ is often replaced by K.

6.3.2.2. *Study of the slow growth phase*

During this phase, nutrients become a limiting factor, cell development decreases, and the rate of BOD oxidation becomes dependent on the BOD concentration in the medium. We write:

$$dX/dt = K\ L\ X$$

Since:

$$\Delta X = \text{am}\ \Delta L$$

It follows that:

$$\frac{dX}{dt} = \text{am}\frac{dL}{dt} \text{ and } \frac{\text{am}}{dX}\frac{dL}{dt} = KL$$

By setting K' = K/am, we obtain:

$$\frac{dL}{dt} = K'XL$$

Assuming an average sludge concentration in tank Xa, the equation becomes:

$$\frac{dL}{dt} = K'XaL$$

This equation is given by integration:

$$\text{Log}\ \frac{Lf}{Lo} = K'$$

Furthermore:

$$L_f = Lo\ \exp\ (-K'\ Xa\ t)$$

Using two values L_1 and L_2 at two fairly distant times t_1 and t_2, the value of K' can be written as:

$$K' = \frac{\text{Log}(L2 - L1)}{Xa(t2 - t1)} \text{ or still } \frac{Lf}{Lo} = \frac{1}{(1 + K'Xat)}$$

This expression is also obtained from the Michaelis–Menten equation. Indeed, the equation linking the substrate evolution as a function of the specific growth rate and the sludge concentration is written as:

$$\frac{dL}{dt} = \mu \frac{L}{K1 + L} X$$

By setting:

$$\frac{dX}{dt} = am \frac{dL}{dt} - bX$$

where b is an endogenous respiration coefficient (d^{-1}).

Combining the two equations gives:

$$\frac{dX}{dt} = am\ \mu \frac{L}{K1 + L} X - bX$$

$$\frac{1}{X}\frac{dX}{dt} = am\ \mu \frac{L}{K1 + L} X - b$$

A term representing the age of the sludge (G) is then defined, which is the inverse of the sludge retention time in the tank. It is expressed in d^{-1}.

In the previous equation, $\frac{1}{X}\frac{dX}{dt}$ is equal to the inverse of the sludge age.

Let:

$$G = 1/t = am\ \mu \frac{L}{K1 + L} X - b$$

At the beginning of treatment, when the substrate is not limiting, that is, it does not limit cell growth, we have L > K1. It can be written that:

$$\frac{dL}{dt} = \mu X$$

At the end of the treatment, when the substrate concentration is low, K1 > L.

The equation becomes:

$$\frac{dL}{dt} = \mu XL/K1$$

$$\frac{dL}{dt} = \frac{\mu X}{K1} dt$$

whose integration leads to:

$$\frac{Lf}{Lo} = \exp(-K'Xt)$$

The series expansion gives:

$$\frac{Lf}{Lo} = \frac{1}{1 + K'Xt}$$

EXAMPLE.– *Study of an industrial effluent.* In a series of beakers, aeration of the effluent is performed in the presence of activated sludge (initial BOD: 1,500 mg.L^{-1}). The results are as presented in Table 6.3.

Beaker	Aeration time (h)	VSS (mg. L^{-1})	Outlet BOD (mg.L^{-1})
1	2	2,000	900
2	4	1,800	570
3	6	1,500	450
4	8	1,300	400

Table 6.3. *Results of BOD_5 degradation in beakers*

Beaker	Xa (g.L^{-1})	Time (h)	X at	Lf (g.L^{-1})	$\frac{Lo-Lf}{Lf}$
1	2	2	4	0.90	0.66
2	1.8	4	7.2	0.57	1.63
3	1.5	6	9	0.45	2.33
4	1.3	8	10.2	0.40	2.75

Table 6.4. *Calculations of the different parameters*

The slope of the line $\frac{Lo-Lf}{Lf}$ with respect to X is equal to 0.327:

$$\frac{Lo - Lf}{Lf} = 0.327\ X\ at$$

6.3.2.3. *Example of determining am*

The determination of am is achieved in the laboratory, following the parallel evolution of VSS and BOD_5. The results obtained are represented in Table 6.5.

Aeration time (h)	BOD_5 (mg.L^{-1})	VSS (mg.L^{-1})
0	1,500	2,000
1	1,140	2,058
2	900	2,108
3	600	2,153
4	500	2,183
5	350	2,196
6	260	2,158
7	190	2,158

Table 6.5. *BOD_5 degradation results*

Based on Monod's law of constant efficiency, we can write that:

$$\Delta X = am\ \Delta L$$

It is assumed that we are in the logarithmic growth phase without taking the endogenous coefficient (b) into account. It can be seen that the first four hours of the test correspond to a phase of logarithmic growth. The determination of am is carried out during this period, by carrying over $\Delta X = f(\Delta L)$.

The calculation leads to a value of am equal to 0.18, with the ratio:

$$am = 180/1{,}000 = 0.18$$

The value of μ can then be determined from the values in Table 6.6.

By carrying over Log $(1 + \frac{am(Lo-L)}{Xa})$ according to t, the value of μ (logarithmic growth phase) can be easily drawn. We obtain $\mu = 0.018\ h^{-1}$.

L ($mg.L^{-1}$)	Lo – L ($mg.L^{-1}$)	$\frac{am(Lo-L)}{Xa}$	$Log\,(1 + \frac{am(Lo-L)}{Xa})$
1,140	360	0.324	0.032
900	600	0.0540	0.052
660	840	0.756	0.720
500	1,000	0.090	0.086
350	1,150	0.103	0.098
260	1,240	0.111	0.105

Table 6.6. *Determination of μ*

6.3.2.4. *Endogenous phase or decay phase*

Since there is only a low substrate concentration, the cells gradually die. This phenomenon can be represented by an exponential law such as:

$$X = X_{max}\ EXP\ (-\ K''t)$$

The endogenous mechanism can be found in all cells where the energy is present to maintain cellular activity. The coefficient b represents the amount of energy consumed daily by the biomass. The values of b differ according to the nature of the substrate, as summarized in Table 6.7.

Discharge	b (d^{-1})
Household	0.07–0.075
Refinery	0.01–0.016
Brewery	0.10–0.12
Paper mills	0.10
Textiles	0.05–0.07

Table 6.7. *Endogenous coefficient b for various types of wastewaters*

6.4. Traditional design methods

Conventional wastewater treatment using the activated sludge process can be performed in two main different ways: complete mixture and piston treatment. There are different design methods for activated sludge that can be found, which in each country and for each manufacturer take into account local conditions and field experience. The symbols grouped in Table 6.8 are those traditionally used in these conventional methods.

Symbols	Function	Units
am Le	Synthesized sludge (new cells and stock)	$kg.d^{-1}$
*Δ*X	Produced biological sludge	$kg.d^{-1}$
BP	Total produced sludge	$kg.d^{-1}$
b Xa	Cell auto-oxidation by endogenous respiration	$kg.d^{-1}$
Cm	Mass load/TSS	$kg\ BOD.kg^{-1}\ TSS.d^{-1}$
C'm	Mass load/VSS	$kg\ BOD.kg^{-1}\ VSS.d^{-1}$
Xb	Excess sludge	$kg.d^{-1}$
BOD	Biological oxygen demand over five days	$mg.L^{-1}$ or $kg.d^{-1}$
COD	Chemical oxygen demand	$mg.L^{-1}$ or $kg.d^{-1}$
Xb	Biological sludge production (bacterial activity)	kg TSS.kg BOD_5
FV	Volatile fraction	%
G	Sludge age	D
GN	Minimum sludge age required for nitrification	D
Le	Removed BOD_5	$kg.d^{-1}$
TSS	Total suspended solids	$mg.L^{-1}$
VSS	Volatile suspended solids	$mg.L^{-1}$
Denitrified N	Denitrified nitrogen	$mg.L^{-1}$ or $kg.d^{-1}$
Nitrified N	Nitrified nitrogen	$mg.L^{-1}$ or $kg.d^{-1}$
O_2	Oxygen requirements	kg of $O_2.d^{-1}$
O_2DN	Amount of oxygen recovered during denitrification	kg of $O_2.d^{-1}$
O_2N	Oxygen requirements for nitrification	kg of $O_2.d^{-1}$
O_2C	Oxygen requirements due to BOD removal	kg of $O_2.kg^{-1}$ BOD_5
O_2N	Oxygen requirements due to nitrogen removal	kg of $O_2.kg^{-1}$ BOD_5
Q	Water flow rate entering the plant	$m^3.d^{-1}$
Si	Flux of hard organic solids in the influent	$kg.d^{-1}$
Xmin	Flux of mineral suspended solids in the influent	$kg.d^{-1}$
Xf	TSS sludge leakage in the effluent	$kg.d^{-1}$
Xa	VSS mass in the aeration tank	Kg
t	Retention time in the aerated tank	h or d
T	Aerobic tank temperature	°C
TV	Total tank volume	m^3

Table 6.8. *Parameters and design units*

In the aeration tank, treatment mechanisms take place:

– physical retention of suspended solids in biological floc;

– physical absorption and then biosorption under enzymatic action;

– oxidation and synthesis of new cells.

An aeration system, which usually supplies the necessary oxygen directly into the biological reactor, is integrated into the system. Once the biological reaction has been completed in the aeration tank, the purified water must then be separated from its purifying biomass. The activated sludge process generally uses the simplest separation technique: clarification (or sedimentation), insofar as the biomass has the consistency of a sludge. The operating conditions of the clarifier should allow for an efficient separation of purified water and biomass. To do this, the hydraulic velocity applied with the unit ($m^3.m^{-2}.h^{-1}$) will have to be smaller than the settling rate of the organic floc. In addition to its function of phase separation, the clarifier, combined with the sludge recirculation, plays the role of biomass concentrator for the aeration tank. The settling tank is used to concentrate the biomass at the bottom of the tank. The recirculation of this concentrated liquor in the biological reactor enables continuous reinoculation of the tank and consequently maintains adequate biological mass.

6.4.1. *Complete mix*

Complete mix is a process that allows settled water to be instantaneously mixed by way of the entire aeration tank. Therefore, there is a constant concentration of activated sludge, homogeneous oxygenation and uniform distribution of organic pollution in the tank. As a result, the applied organic load is uniform, being ensured by the dispersion of the effluent. We write that the removed pollution is proportional to the sludge mass existing in the aerator and to the substrate concentration.

The process is applied for direct raw water treatment or downstream of a primary settling stage. Let:

– Q denote the treated water flow rate;

– Qr denote the sludge recirculation rate;

– Xa denote the biomass concentration (activated sludge) in the aeration tank;

– Xr denote the biomass concentration at the bottom of the clarifier and in the recirculation (return sludge).

6.4.1.1. *Retention time (rt)*

The retention time represents the average contact time in the aeration tank for the purifying biomass and wastewater. It is expressed in hours or days and defined from the average operating conditions.

Hydraulic retention time:

$$t = V/Q \text{ (h)}$$

where V is the reactor volume in m^3, and Q is the wastewater supply flow in $m^3.h^{-1}$.

6.4.1.2. *Applied loading*

Generally speaking, it is not easy to know the quantity of active biomass in a sludge, and it is only expressed by the weight of the VSS. The characteristics of a reactor are defined by two factors:

– the volumetric loading;

– the mass loading.

6.4.1.2.1. Volumetric loading (Cv)

When a certain weight of organic matter expressed in BOD_5 (kg/day) has to be transformed daily in an aerobic reactor of volume V (m^3), the volumetric loading (Cv) is defined as the quantity of pollution admissible into the aeration tank per unit volume and per unit of time. It is expressed in $kg.m^{-3}.d^{-1}$ from the average polluting flux:

$$\mathrm{Cv} = \frac{\text{Daily pollutant flux}}{\mathrm{V}} = \frac{\mathrm{Q\,L}}{\mathrm{V}}$$

where L is the pollution concentration in the admissible raw water for treatment in $kg.m^{-3}$ or $g.L^{-1}$ (generally expressed in BOD).

6.4.1.2.2. Mass loading (Cm)

When maintaining a quantity of activated sludge (Xa) in the tank, we then define the mass loading (Cm), which quantifies the substrate/biomass ratio in the aeration tank. It is expressed in $kg.kg^{-1}.d^{-1}$ for average operating conditions in a steady state:

$$\mathrm{Cm} = \frac{\text{Daily pollutant flux}}{\text{Quantity of biomass in the reactor}} = \frac{\mathrm{Q\,L}}{\mathrm{VXa}}$$

where V Xa is the quantity of biomass (activated sludge) present in the reactor (kg), and Xa is the biomass concentration in the reactor ($kg.m^{-3}$).

It is generally estimated by the total suspended solids (TSS) or VSS. These parameters represent only an estimate of biomass insofar as they do not differentiate between living matter and inert matter.

Mass loading is traditionally expressed in kg $BOD.kg^{-1}$ $TSS.d^{-1}$ or kg $BOD.kg^{-1}$ $VSS.d^{-1}$.

Mass loading C'm (in kg BOD/kg VSS.d) and transit time (in h) are related by the equation:

$$C'm = \frac{\text{inlet DBO Q}}{\text{Xa V}} = \frac{\text{inlet DBO 24}}{\text{Xa t 1000}}$$

The applied mass and volumetric loadings define the type of treatment. They are valid in temperate climatic conditions (T = 10 to 20 °C).

Plants that involve activated sludge are classified according to their level of applied mass loading. The lower it is, the more thorough the substrate consumption and therefore pollution removal. Depending on the applied loading conditions, the following orders of magnitude are obtained for urban wastewater, both raw water and settled water, for the BOD removal efficiency:

– low loading: Cm < 0.15 kg $BOD.kg^{-1}$ $SS.d^{-1}$;

– medium loading: 0.15 < Cm < 0.7;

– high loading: Cm > 1.

Since mass loading and sludge age inversely evolve, low loadings correspond to high sludge ages and make it possible to select a biology with a low growth rate. In contrast, a low sludge age will be obtained under high-loading conditions where only species with a high growth rate are able to survive.

Only low loadings make it possible to develop a biology capable of eliminating nitrogen pollution. At the same time, they allow high removal efficiencies with biodegradable organic pollution.

Extended aeration is a special case of low-loading processes. This is the only way to obtain stabilized sludge in parallel with the water treatment.

Medium loadings together with settled domestic wastewater enable obtaining, besides nitrogen removal, efficient results with organic pollution, and make it possible to restore a treated effluent that complies with the former "e" level of French legislation or the "non-sensitive zone" level of the European directive. The upper end of the scale is only applicable with very diluted water.

High loadings correspond to conditions that favor phenomena of biosorption and absorption of particulate pollutants in the floc.

This scale varies according to the authors. Some recommend not splitting the low load of the extended aeration, extending the range from 0.05 to 0.2 kg $BOD_5.kg^{-1}$ $VSS.d^{-1}$.

Others do not mention very high loadings, only admitting a scale of high loadings ranging from 0.8 to 4. Table 6.9 presents the values of mass and volumetric loadings according to the mode of operation of the biological tank.

	Extended aeration	**Low loading**	**Medium loading**	**High loading**	**Very high loading**
Cm kg $BOD.kg^{-1}$ $TSS.d^{-1}$	0.04 to 0.08	0.08 to 0.15	0.25 to 0.75	0.75 to 1	> 1
C'm kg $BOD.kg^{-1}$ $VSS.d^{-1}$	0.06 to 0.1	0.1 to 0.2	0.33 to 1.1	1.1 to 1.4	> 1.4
Cv kg $BOD.m^{-3}.d^{-1}$	0.2 to 0.32	0.32 to 0.6	0.8 to 1	1 to 2	> 2

Table 6.9. *Cm, C'm and Cv according to the operating mode of the biological tank*

The SS and VSS concentration values in tanks are presented in Table 6.10.

Treatment systems	**X/(SS) Sludge mass ($kg.m^{-3}$ SS)**	**X/(VSS) Sludge mass ($kg.m^{-3}$ VSS)**	**Mass loading (kg $BOD.kg^{-1}$ $VSS.d^{-1}$)**	**Volumetric loading (kg $BOD.m^{-3}.d^{-1}$)**	**% BOD_5 removal**
Extended aeration	5–6	4	0.1	0.3	98
Low loading	5	3.5–4	0.1–0.2	0.3–0.8	95
Medium loading	4	3–3.5	0.2–0.6	0.8–1.8	90
High loading	3	2.5	0.9	1.8	85

Table 6.10. *X/(SS) and X/(VSS) according to the operating mode of the biological tank*

The relation indicates the mass loading as a function of the BOD efficiency obtained (% removed BOD_5):

$$\mathrm{Cm} \leq \left(\frac{99.5 - \mathrm{Effic}\,\%}{15}\right)$$

6.4.1.3. *Sludge age (G)*

The sludge age G represents the average retention time of the sludge in the aeration tank and is expressed in days. The removal of pollution induces the production of biomass in the form of sludge (produced sludge):

$$G = \frac{X}{\Delta X}$$

with ΔX being the sludge increase (produced and extracted from the system in steady state) per unit of time, in $kg.d^{-1}$.

The simplified relation is:

$$G = \frac{Xt}{\Delta X} = \frac{Xt}{amLe - bXa}$$

where Xt = XaV represents the total mass of sludge in the biological tank (kg), and Xa is the sludge concentration in the biological tank ($kg.m^{-3}$).

Sludge age is the dominant parameter of biological treatment, the one that defines the biological balance of the biomass and its treatment efficiency. For a given substrate, it varies inversely to the mass load parameter Cm:

$$G = \frac{1}{amCm - b}$$

		Extended aeration	Low loading	Medium loading	High loading	Very high loading
Sludge age G (days)		18 to 30 (20 recommended)	13 to 18 (15 recommended)	3 to 7 (4 recommended)	1 to 3 (2 recommended)	< 1 (0.5 recommended)
Efficiencies	C	95%	90%	90%	80%	50%
Removal	N	95%	90%	–	–	–
Indicative	P	Yes	Yes	Yes	–	–

Table 6.11. *Sludge age (G) according to the operating mode of the biological tank*

For a sludge age G to be maintained in the treatment plant in order to meet the treatment and quality conditions of the outlet effluent, the following values are recommended:

G = 2 d to obtain a 30 $mg.L^{-1}$ BOD_5 concentration at the plant outlet;

G = 4 d to obtain a 15 mg.L^{-1} BOD_5 concentration at the plant outlet;

G ≥ 13 d at 12 °C to obtain a 15 mg.L^{-1} NGL concentration at the plant outlet;

G ≥ 18 days at 12 °C to obtain a 10 mg.L^{-1} NGL concentration at the plant outlet.

6.4.1.4. *Organic pollution removal (efficiency law)*

In the established regime and neglecting the leakage of suspended solids into the effluent, the mass conservation of suspended solids in the system results in the following equilibrium:

$$(Q + Qr).Xa = Qr.Xr$$

$$Xa = \frac{QrXr}{Q+Qr}$$

The pollution removed daily is Q (Lo – Lf) or QLe, and that remaining in the tank is V Lf (kg). By denoting with M the fraction of substrate consumed for metabolism, it follows that:

$$Q\ (Lo - Lf) = M\ V\ Lf$$

Dividing by Q, and setting V/Q = t, we have:

$$Lo - Lf = M\ t\ Lf$$

Given that the complete mix implies that the water to be treated is intimately mixed with a homogeneous mass of activated sludge present in the tank, and that this interacts with the time parameter in the form of the monomial Xa t, it follows that:

$$\frac{Lo - Lf}{Lf} = K\ Xat$$

This equation is equivalent to the previous equation:

$$\frac{\text{outlet soluble BOD-}}{\text{inlet BOD}} = \frac{Lf}{Lo} = \frac{1}{1 + KXa\ t}$$

This equation can therefore be applied to a complete mix and when referring to a phase of slow growth.

6.4.1.4.1. Alternative method

The problem can be approached in another way by considering recycling. It can be said that complete-mix reactors with no recycling suitably simulate anaerobic digestion or even aerobic and anaerobic ponds.

On the contrary, complete-mix reactors with recycling simulate activated sludge treatment (Figure 6.3).

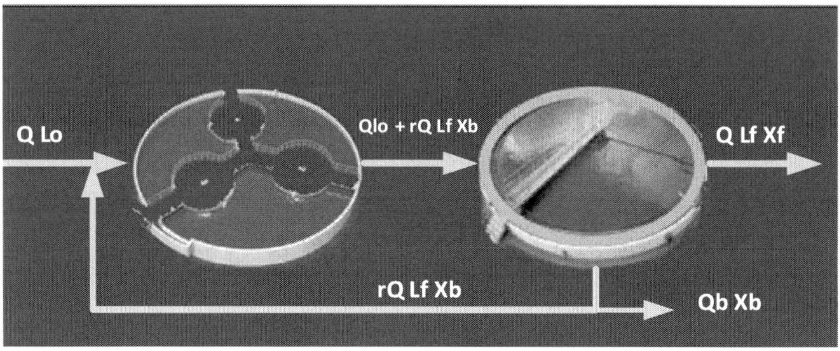

Figure 6.3. *Complete mix with recycling*

In steady state, the sludge balance is written as:

$$V\frac{dX}{dt} = am\frac{dL}{dt}V - bXaV$$

with:

– dX/dt being the variation in the VSS concentration in the reactor;

– V being the reactor volume;

– dL/dt being the change in BOD_5 concentration in the reactor dt.

By dividing by X, the sludge age (G) can then be defined:

$$\frac{1}{X}\frac{dX}{dt}V = \frac{am}{X}\frac{(-dL)}{dt}V - bV = \frac{V}{G}$$

or:

$$\frac{1}{G} = \frac{am}{X}\frac{(-dL)}{dt} - b$$

We know that the BOD variation is expressed as:

$$\frac{dL}{dt} = \frac{\mu\, X\, L}{K + L} = \frac{\mu\, X\, Lf}{K + Lf}$$

$$\frac{1}{G} = \frac{am\,\mu\,Lf}{K + Lf} - b$$

Therefrom, it can be derived that:

$$Lf = \frac{(K + bG)}{G(am\,\mu - b) - 1}$$

The concentration Lf is therefore independent of the inlet concentration Lo. A variation in Lo is reflected by a variation in X, allowing Lf to be kept at the desired value.

During a finite time interval, it can be written that:

$$\frac{dL}{dt} = \frac{Lo - Lf}{t}$$

with t being the water retention time in the reactor, equal to: t = V/Q.

Inserting t into the sludge age equation gives:

$$\frac{1}{G} = \frac{am\,(Lo - Lf)}{Xa\,t} - b$$

$$Xa = \frac{am\,(Lo - Lf)}{1 + bG}\,\frac{G}{t}$$

When there is no recycling, the retention time is equal to the sludge age:

$$G = t$$

The previous equations then become:

$$Xa = \frac{am\,(Lo - Lf)}{1 + bt}$$

$$Lf = \frac{K\,(1 + bt)}{t(am\mu - b) - 1}$$

The biomass concentration in the tank can be characterized with the VSS parameter, the relation then generally being expressed as $K = 0.5h^{-1}g^{-1}$ when taken at low loading and K = 0.8 at 1.4 $h^{-1}g^{-1}$ at average load, with t = water retention time in the tank (in hours) and Xa = VSS % concentration in the tank ($g.L^{-1}$ or $kg.m^{-3}$).

The relation can be expressed with respect to the mass loading:

$$\text{outlet soluble BOD} = \text{Lf} = \frac{\text{inlet BOD}}{1 + \text{K}\,\dfrac{\text{inlet BOD 24}}{\text{C'm 1,000}}}$$

BOD is expressed in mg.L^{-1}, and the mass load in kg BOD.kg^{-1} VSS.d^{-1}:

$$\text{BOD efficiency} = 1 - \frac{\text{outlet soluble BOD}}{\text{inlet BOD}} = \frac{1}{1+\dfrac{1{,}000\ \text{C'm}}{\text{inlet K BOD 24}}}$$

Quite often with this conventional method, soluble BOD effluent is associated with total BOD effluent. In some cases, a particulate BOD leak into the effluent may be taken into account:

Effluent total BOD = effluent soluble BOD + effluent particulate BOD

The effluent particulate BOD is estimated for an extended aeration generally at 0.3 mg BOD_5 per mg of suspended solids; a value of 0.5 mg BOD/mg TSS is taken for a medium-loading operation. This is the only approach for defining the effectiveness of the treatment for organic pollution. COD, another standardized parameter of organic pollution and for which discharge standards are systematically requested, is therefore never taken into account.

6.4.1.5. *Sludge production*

Organic matter removal leads to cell synthesis that increases the proportion of sludge in the pond. It is then necessary to know the mass of formed sludge in order to predict the volume of discharges and the treatment processes for the extracted sludge.

The evaluation of the surplus sludge balance can be estimated by the difference between the formed or added sludge and the removed sludge. The increase in sludge in a tank is the result of various phenomena such as bacterial growth, with an efficiency am and possible reserves Pe, as the solids are brought by the effluent. These will be mineral or organic.

Mineral materials that are not affected by biological activity are called X min.

The effluent to be treated is also made up of organic matter that does not biodegrade easily due to its chemical nature (size, substituted products, acidity, etc.). They are generally referred to as “hard”. They represent a small fraction of biodegradable materials.

Sludge reduction in a tank results from endogenous respiration, which leaves a residue estimated at about 0.20 b Xt, and from the SS leakage with the clarified effluent X eff, which depends on the current discharge standards. A value of 30 $mg.L^{-1}$ is in general adopted.

The activated sludge treatment therefore generates sludge (produced sludge) and consumes oxygen. Depending on the retention time of the biology in the system (sludge age), the auto-oxidation reaction will be more or less intensive. The overall result of the process leads to:

– for a high sludge age and low mass loading, high oxygen consumption and reduced sludge production;

– for a low sludge age and a high mass loading, reduced oxygen consumption but increased sludge production.

In terms of produced sludge, in addition to the biological result, we have:

– an inert fraction in the biological treatment, trapped however in the biological floc and comprising of suspended mineral matter and suspended inert organic matter brought by the wastewater;

– a biodegradable fraction that is biosorbed or absorbed on the floc and has not been hydrolyzed. The older the sludge, the smaller this fraction will be.

The proportion of these various components in the produced sludge is related to the characteristics of the wastewater and the operating conditions of the activated sludge.

– Relation between sludge age, mass loading and specific sludge production:

$$G = \frac{X}{\Delta X}$$

$$Cm = \frac{Q\ BOD}{VX}$$

$$GCm = \frac{X}{\Delta X}\frac{Q\ BOD}{VX} = \frac{Q\ BOD}{X}$$

– Production of biological sludge originating from BOD removal: this is the result of two parameters, one proportional to the removed pollution Le (expressed in BOD), the other to the quantity of biomass present in the system Xa bio (expressed in VSS):

$$\Delta X \text{ (expressed in VSS)} = am\ Le - b\ Xa$$

The first term represents the instantaneous production of VSS resulting from BOD degradation, and the second term indicates that part of the instantaneous production disappears (corresponding to the endogenous auto-oxidation of the biomass). The second term never cancels the first one; with very long sludge retention times in the system (high sludge age, "total oxidation"), the resultant tends to a value corresponding to 0.25 Le.

6.4.1.6. *Total sludge production*

For an influent that is not exclusively soluble and therefore contains a fraction of the pollution in particulate form, in addition to the biological production previously defined, there is also an inert production provided by the influent. The total production is then written as:

$$\text{Total } \Delta X = X\text{min} + Si + \Delta X$$

$$\text{Total } \Delta X = X\text{min} + Si + am\ Le - b\ Xa$$

with:

– Total ΔX being the quantity of produced sludge (kgTSS.d^{-1});

– Xmin being the mineral suspended solids flux entering with the influent estimated in general at 33% of the TSS;

– Si being the flux of hard organic solids (VSS) brought in by the influent and generally estimated at 25 to 30% of the incoming VSS (more or less high depending on the biological loading conditions);

– am being the biological sludge synthesis coefficient (kgVSS.kg^{-1} removed BOD);

– Le being the flux of removed BOD (kg.d^{-1});

– b being the biomass auto-oxidation coefficient (d^{-1});

– Xa being the amount of biomass present in the system (kgVSS).

Since in such a system it is not possible to differentiate between biological and other VSS, the previous relation is extrapolated with the use of the total mass of VSS in the system with the relation:

$$\text{total } \Delta X = X\text{min} + Si + am\ Le - b\ Xa$$

with Xa being the total amount of VSS (= Xa V in kg) present in the activated sludge tank.

Under these conditions, and for a purely domestic influent, the following values are generally used: am = 0.65 $kgVSS.kg^{-1}$ BOD in aeration and 0.55 to 0.6 $kgVSS.kg^{-1}$ BOD at medium loading; b = 0.07 d^{-1} in prolonged aeration and 0.08 to 0.1 d^{-1} under medium load.

6.4.1.7. *Excess sludge*

Depending on the efficiency of the clarification process, the sludge produced is discharged from the system, partly by the leaking of TSS with the treated effluent and partly in a voluntary manner by the operator through the extraction of excess sludge. This second fraction will be the only one subjected to sludge treatment.

– Excess sludge = produced sludge – TSS leakage:

$$\text{Xe} = \text{total } \Delta\text{X} - \text{Xf}$$

Xf = effluent TSS leakage

$$\text{Xe} = \text{Xin} + \text{Si} + \text{am Le} - \text{b Xa} - \text{Xf}$$

– Relation with G, the sludge age:

$$\text{G} = \frac{\text{Xa}}{\Delta\text{X}} = \frac{\text{Xa}}{\text{Xmin} + \text{Si} + \text{amLe} - \text{bXa}}$$

where G is the sludge age (j), and Xa is the total TSS mass in the activated sludge tanks (Xa V in kg).

Considering that, in this approach, the only mineral component in the sludge is due to Xmin, the relation with sludge age can be expressed with respect to the VSS of the system:

$$\text{G} = \frac{\text{Xa}}{\Delta\text{X}} = \frac{\text{Xa}}{\text{Si} + \text{amLe} - \text{bXa}}$$

The general formula for sludge production can then be represented as:

$$\Delta\text{X} = \text{Xmin} + \text{Si} + \frac{\text{am Le} - \text{b G Si}}{1 + \text{bG}}$$

The total masses of sludge in the tanks are:

$$\text{Xa} = \text{G}\,(\text{Xmin} + \text{Si} + \frac{\text{am Le} - \text{b G Si}}{1 + \text{bG}})$$

with Xa in kg TSS.

Expressed in total VSS in the tanks, the biomass concentration in the reactor is obtained by:

$$Xa = G\,Si + G\frac{am\,Le - b\,G\,Si}{1 + bG}$$

The first term represents the mass of inert VSS brought by the tributary and accumulated in the sludge; the second term quantifies the mass of biological sludge created by the BOD degradation.

For a given effluent, sludge production is low for a mass load, which is itself lower. It is assumed that this production is minimal when the term (am, Le – b, Xt) is close to 25% of the living matter initially produced:

$$am\ Le - b\ Xt = 0.25\ am\ Le$$

In fact, this has only been observed for mass loadings of less than 0.1 kg $BOD_5.kg^{-1}$ $VSS.d^{-1}$. This is then known as “total oxidation”, also called “extended aeration”.

The application of this formula results in estimating that 40 to 45 $g.d^{-1}$ is the amount of sludge produced by extended aeration treatment of an urban effluent, assuming a discharge of 60 g BOD_5 per capita per day and approximately 70 g of $SS.d^{-1}$.

6.4.1.8. *Recirculation dimensioning*

Calculated is undertaken in different ways using the sludge volume index (SVI).

6.4.1.8.1. From the Mohlman index (SI or MI or SVI)

The SVI expresses the settling ability. For the same quality of sludge, SI varies according to the sludge concentration. This index is measured with a one-liter sample of freshly collected sludge:

$$SI\ (ml.g^{-1}) = V_{30}/Xa$$

with:

– V_{30} being the volume of the sludge blanket after 30 minutes of settling ($ml.L^{-1}$);

– Xa being the TSS concentration in the tested sample of activated sludge ($g.L^{-1}$).

When V_{30} is greater than 250 ml, the measurement being influenced by the diameter of the specimen, it is essential to carry out a dilution (V_{30} after dilution = 150 to 250 ml.L^{-1}). The concentration used in the calculation will be that of the diluted sample. These criteria are given for a temperature between 12 and 15 °C.

The maximum concentration of return sludge is given by:

$$Xr\ (g.L^{-1}) = 1{,}200/MI$$

with:

– Xr being the maximal concentration of return sludge;

– MI being the Mohlman index.

The minimum percentage that has to be ensured depends on the VSS concentration in the aerator and the MI, such as:

$$r = \frac{100Xa}{\frac{1{,}200}{MI} - Xa}$$

with:

– r being the return percentage of sludge to be recycled;

– Xa being the average concentration of VSS in the aerator (g.L^{-1});

– MI being the Mohlman index.

Table 6.12 gives the different values of r for variable VSS concentrations and an MI:

$$(Qo + Qr)\ Xa = Qr\ Xr + Qbe$$

$$r = Xr/Qo$$

with:

– Qo being the flow rate of water to be treated;

– Qr being the recirculation rate of the sludge originating from the clarifier;

– Qbe being the excess sludge flow;

– Xa being the TSS concentration in the biological tank;

– Xr being the TSS concentration in the return sludge;

– r being the recirculation rate (%).

The recirculation rate is usually set at 100% of the instantaneous flow rate. Recirculation pumps are dimensioned such that they recirculate 200% of the rated hourly flow rate. It is also necessary to ensure that the maximum flow rate of the pumping plant is greater than 67% of the hourly peak rainy weather flow.

Xa **VSS (g.L^{-1})**			**r (%)** **MI**		
–	100	150	200	250	300
1	9	14	20	27	33
1.5	14	23	33	46	60
2	20	33	50	73	100
2.5	26	45	72	110	168
3	33	60	100	171	300
4	50	100	200	500	–
5	71	166	500	–	–

Table 6.12. *Recirculation rate according to the MI and biomass concentration*

A variable MI between 100 and 150 is considered satisfactory and suggests a good settling of the sludge. It is obvious that for higher values, settling may become an issue. The problem can also be solved from the reference volume. Indeed, the volume occupied by sludge Xa will be equal to:

Vb = MI Xa (expressed in ml.L^{-1})

For example, a sludge with an index of 100 will be limited to a maximal sludge concentration equal to 10 g.L^{-1}, since the volume occupied by the sludge will be:

Vb = 100 x 10 = 1,000 ml.L^{-1}

This volume can also be deduced from the following relation:

$$Vb = \frac{1{,}000\ r}{Q + r}$$

where r is the recycling flow rate, and Q is the inlet flow rate into the plant.

This relation can take the form of:

$$\frac{Vb}{1{,}000} = \frac{r}{Q + r}$$

or:

$$\frac{1,000}{Vb} = \frac{Q}{r} + 1$$

That is:

$$\frac{r}{Q} = \frac{Vb}{1,000 - Vb}$$

And finally:

$$\frac{r}{Q} = \frac{MI\,Xa}{1,000 - MI\,Xa}$$

6.4.1.8.2. From the Kalbskopf index

The recirculation rate will depend on the Kalbskopf index (KI), which can be estimated from the experimental curve: concentration (VSS) during recycling according to the KI.

A compaction rate rc is then defined, which is equal to:

$$rc = \frac{Xr}{Xa}$$

with:

– Xr being the recycled sludge concentration ($g.L^{-1}$);

– Xa being the sludge concentration in the aerator ($g.L^{-1}$).

The recycling rate becomes:

$$r = \frac{1}{rc - 1}$$

The concentration of recycled sludge can be directly estimated by the following relation:

$$Xr = \frac{1,200}{MI} \text{ in } g.L^{-1}$$

6.4.1.8.3. From the Donaldson Index (DI)

This density index is obtained with respect to the volume of settled sludge (expressed in %) in half an hour:

$$DI = \frac{Xa}{100 \text{ x settled volume } (\%)}$$

or still:

$$\text{settled volume } (\%) = \frac{Xa}{100 \text{ DI}}$$

It is estimated that the maximum volume of sludge allowed during settling is 20%. As an indication, Table 6.13 presents the different density indexes for variable VSS concentrations.

Xa ($mg.L^{-1}$)	DI
800	0.4
100	0.5
1,500	0.75
2,000	1

Table 6.13. *Donaldson index according to biomass concentration*

In a plant with several parallel basins, the recirculation of mixed liquors and return sludge will be independently regulated in each basin, in order to facilitate intervention if bulking occurs.

In the event of a control failure, two types of emergency controls are possible for the recycling of mixed liquors and sludge produced by clarifiers:

– constant flow recirculation calculated from the average daily flow rate of water to be treated;

– control according to a timetable if it is already implemented for another control. The timetable should give the operating frequencies of the pumps or the number of pumps operating during each period.

EXAMPLE.– What is the recycling rate that should be expected in an activated sludge system, if the VSS concentration in the tank is equal to 2,000 $mg.L^{-1}$ and the MI equal to 100?

– From MI:

$$\text{recycling rate r} = \frac{2}{\frac{1{,}200}{100} - 2}$$

r = 20% of average flow

– From KI: if this index is estimated at 100, the recycled sludge concentration is about 12 $g.L^{-1}$.

The compaction rate is:

rc = 12/2 = 6

The recycling rate is:

$$r = \frac{1}{6-1} = 20\% \text{ of average flow}$$

The sludge concentration in the tank is deduced from the relation:

Xr = 1,200/MI = 1,200/100 = 12 $g.L^{-1}$

which leads to the same recycling rate.

6.4.2. *Plug flow*

This process is also called "with-the-flow" treatment. In this case, with the exception of longitudinal diffusion, the effluent travels in blocks. A constant variation is no longer seen in dL/dt since BOD varies in the reactor. It is written that:

$$\frac{\Delta L}{\Delta t} = \frac{Lo - Lf}{t}$$

$$\frac{1}{G} = \frac{am\ (Lo - Lf)}{X'a\ t} - b$$

with a sludge mass X'a in the reactor.

We take the integral of dL/dt between Lm and Lf where Lm is the average BOD_5 obtained after recycling. It then follows that:

$$Lm = \frac{QLo + rLf}{Q + r}$$

When the retention time varies from t = 0 to t = V/(Q + r):

$$(Lm - Lf) + K1\ Log \frac{Lm}{Lf} = \mu\ X'a\ t$$

Replacing Lm by its value yields:

$$\left(\frac{QLo+rLf}{Q+r} - Lf\right) + K1\,Log\left(\frac{\frac{QLo+rLf}{Q+r}}{Lf}\right) = \mu\, X'a\, t$$

By integrating X'a into the relation of the sludge age, we have:

$$\frac{1}{G} = \frac{am\ \mu\ (Lo - Lf)}{(Lo - Lf)r\left(1 + \frac{r}{Q}\right) Log(\frac{r}{Q} Lf + Lo)/(1 + \frac{r}{Q})Lf\ K1}$$

with:

$$X'a = \frac{am\ (Lo-Lf)}{1+bG}\frac{G}{t}$$

A sludge balance around the reactor leads to: return = outlet – extraction. That is:

$$r\ Xr = (Q + r)\ X'a - r'\ X'a$$

By setting:

$$G = \frac{VX'a}{r'Xr+(Q-r')Xd}$$

the term (Q-r') Xd corresponds to the sludge leakage with the effluent. It can be neglected if the settler is properly sized. It then remains that:

$$G = \frac{VX'a}{r'Xr}$$

where X'a represents the sludge concentration in the reactor, and Xr represents the recycled sludge concentration.

Since recycled sludge and settled sludge have the same concentration:

$$(Q + r)\ X' = r\ Xr + r'\ Xr$$

$$\frac{1}{G} = \frac{r'Xr}{VX'a}$$

Replacing r′ Xr with its value, it follows that:

$$\frac{1}{G} = \frac{(Q + r)X'a - rXr}{VX'a} = \frac{Q}{V} + \frac{r}{V} - \frac{rXr}{VX'a}$$

Since:

$$t = V/Q$$

we obtain:

$$\frac{1}{G} = \frac{1}{t}(1 + \frac{r}{Q} - \frac{rXr}{QX'a})$$

This indeed leads to t = G when there is no recycling.

6.4.3. *Comparison between the two systems*

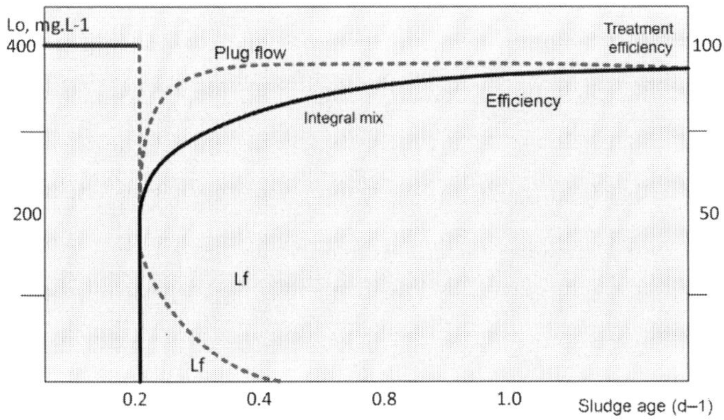

Figure 6.4. *Comparison of the obtained efficiencies in the two processes*

Since the reactor operates in plug flow (oxidation channel), it allows for higher BOD removal returns compared to the complete-mix reactor. This is achieved for shorter retention times. Nonetheless, the complete-mix system is preferred for its stability and the low variations in the concentration of treated effluent. The instantaneous dilution of the raw effluent in the reactor facilitates absorbing sudden changes in the load or even the feeding of toxic substances.

6.4.4. *Oxygen requirements for carbon pollution*

Sludge oxygen consumption results from a superposition of catabolic phenomena (energy production for synthesis) and cellular breathing phenomena. Various studies have made it possible to specify a symbolic formula for the organic matter contained in an urban effluent. It is $C_7 H_{11} NO_3$.

As far as living matter is concerned, two formulas are proposed: ($C_5 H_7 NO_2$) or ($C_7 H_9 NO_3$).

The overall catabolism reaction is written as:

– For an urban effluent:

$$C_7H_{11}NO_3 + 15/2\ O_2 \rightarrow 7CO_2 + NH_3 + 4\ H_2O$$

– For a particular substrate such as glucose, for example:

$$C_6H_{12}0_6 + 60_2 \rightarrow 6CO_2 + 6H_2O$$

The synthesis reaction is written as:

– For glucose:

$$6\ C_6H_{12}0_6 + 160_2 + 4\ NH_3 \rightarrow 4\ C_5H_7NO_2 + 16\ CO_2 + 28\ H_2O$$

Therefore, 1,080 g of glucose would enable 452 g of cells to be synthesized:

– For an urban effluent:

$$C_7\ H_{11}NO_3 + \tfrac{1}{2}\ O_2 \rightarrow C_7H_9NO_3 + H_2O$$

Generally speaking, the substrate is estimated in BOD or COD. The overall response for cellular respiration is:

$$C_5H_7NO_2 + 5O_2 \rightarrow 5CO_2 + NH_3 + 2H_2O$$

It can be assumed that the oxygen consumption per unit of time is the sum of four terms corresponding to:

– the degraded substrate: p mc Le;

– cell synthesis: q ms Le;

– endogenous respiration: b' Xt;

– the use of any stocks: (p me + q ms) Δ Ps.

The overall equation then becomes:

$$q0_2/t = (p\ mc + q\ ms)\ (Le + \Delta Ps) + b'\ Xt$$

where p and q are respectively the weight of oxygen used for the oxidation of organic matter and for synthesis. p is about 1.52 for an average effluent and 1.06 for glucose following the oxidation reaction:

$$C_6\,H_{12}\,O_6 + 6\,O_2 \rightarrow 6\,CO_2 + 6\,H_2O$$

$$p = 192/180 = 1.06$$

The value of q is about 0.10 for $C_7\,H_{11}\,NO_3$ and 0.47 for glucose.

The ratio between the weight of removed BOD (Le, Kg BOD_5 removed) and the total admissible pollution (Pe, Kg BOD_5 inlet) is represented by f. That is:

$$f = Le/Pe$$

By letting:

$$a' = \frac{pmc+qma}{f}$$

the general equation remains as:

$$qO_2/t = a'\,(Le + \Delta Ps) + b'\,Xa$$

When the reserves are significant, there is no stock consumption ($\Delta Ps = 0$). The requirements in O_2 correspond to the sum of two terms and are expressed as:

$$qO_2/t = a'\,Le + b'\,Xa$$

The first term proportional to the flow of removed BOD corresponds to the oxygen needed to transform the pollution into biological cells (synthesis). The second term proportional to the mass of VSS in the tanks (representing the existing biomass) corresponds to the oxygen used for sludge auto-oxidation (endogenous respiration):

– a′ is the quantity of oxygen needed to synthesize one kg of fed BOD_5 = 0.65 kg $O_2.kg^{-1}$ of BOD_5 in extended aeration, 0.55 at average load;

– Le is the quantity of BOD5 to be removed during the considered period in $kg.d^{-1}$;

– b′ is the daily quantity of oxygen necessary for the endogenous metabolism of 1 kg of volatile suspended matter (VSS) in the sludge; $b' = 0.07\ kgO_2.kg^{-1}\ VSS.d^{-1}$ in extended aeration, 0.08 to 0.1 at medium load;

– Xa is the VSS mass in the aeration tank (kg).

The values of a' vary between 0.5 and 0.65. They depend on the applied load (Table 6.14).

Mass loading ($kgBDO_5.kg^{-1}$ $VSS.d^{-1}$)	0.90	0.10	0.15	0.20	0.30	0.40	0.50
a'	0.66	0.65	0.63	0.59	0.56	0.53	0.50

Table 6.14. *Values of a' according to the applied mass load*

With a' we therefore denote the fraction of pollution transformed into synthetic energy during the treatment. It is also the amount of oxygen to be supplied to the bacteria so that they synthesize living matter from 1g of pollution.

When the feed becomes limited to satisfy the needs of the cells, stocks are then consumed. As a result, the amount of oxygen becomes equal to:

$$qO_2/t = a' (Le + Ps) + b' Xt$$

If Le + Ps = 0, after utilization of the reserves, there remains:

$$qO_2/t = b' Xt$$

In practice, during wastewater treatment, it is assumed that stocks accumulated at peak hours will be consumed at off-peak hours, allowing for a feed balance during the day. The overall formula becomes:

$$qO_2/t = a' Le + b' Xt$$

Since the cell efficiency am expresses the amount of formed new cells (mg) per mg of removed BOD and cell oxidation with the formula ($C_5H_7NO_2$) requires 1.42 times its oxygen weight, it is considered that 1.42 am becomes the BOD fraction used for the synthesis of these new cells. We define that a = 1.42 am.

To synthesize these cells, it is necessary beforehand to feed a certain amount of energy originating from the oxidation of a certain quantity of organic matter, thus providing a quantity of oxygen, which is the supplement: for a ratio COD/BOD_5 = 1.46 (wastewater), we then have 1.46 = a + a′, with a representing the fraction of pollution transformed into living matter for bacterial synthesis. The maximal value of a is obtained for an extended aeration treatment, that is, when the feed is low.

For a value of a' equal to 0.66 (Cm: 0.09), it follows that:

$$a = 1.46 - 0.66 = 0.80$$

When the material balance is established according to the COD or ultimate BOD, we obtain:

$$a + a' = 1$$

This relation derives from the fact that it is all the organic carbon present in the substrate that leads to the production of CO_2 or the formation of new cells. In this case, $1 - 1.42\ am = 1 - a = a'$, that is, $a + a' = 1$.

For endogenous respiration, the previous equation relating to the oxidation of living matter was written as:

$$C_5H_7NO_2 + 5O_2 \rightarrow 5CO_2 + NH_3 + 2H_2O$$

Therefore, 113 g of living matter requires 160 g of O_2 to be mineralized.

If we call b the coefficient representing the quantity of volatile matter destroyed by endogenesis to provide the maintenance energy, we have:

$$b' = \frac{160}{113} b \text{ that is, } b' = 1.42\ b$$

where b is expressed in g destroyed $VSS.g^{-1}$ sludge $VSS.d^{-1}$; b′ is expressed as g $O_2.g^{-1}$ $VSS.d^{-1}$.

In the general equation of oxygen consumption, qO_2/t is expressed in $kg.d^{-1}$, requiring that the removed BOD be expressed in $kg.d^{-1}$ and that Xt representing the total mass of sludge present in the aerator be reported in kg. The values of coefficients a' and b' are obtained in the laboratory using a respirometer.

6.4.4.1. *Measurement of coefficients a' and b'*

There are different devices for measuring these coefficients, which are classified into two categories:

– constant pressure respirometers;

– constant volume respirometers.

6.4.4.1.1. Constant pressure respirometers

Gilson device

The most widespread device is the Gilson device, which is based on a measurement unit that consists of a measuring vial, a pressure gauge, a display for the volume and also a tap that differentiates the measuring vial from the reference vial. The device also includes valves that allow the unit of measurement to be switched on or off or to admit the gas with which the experiment will be carried out. Gas exchange causes a decrease in volume in a vial and results in a change in the index. After measuring, it is readjusted by a compensation screw, reflecting the loss of volume due to the disappearance of oxygen. The measurements are carried out at a constant temperature (20°C), by immersing the measuring vials in a thermostatic bath. All tests are carried out at constant pressure, which is the atmospheric pressure at the time.

Sierp device

This respirometer makes it possible to keep the liquid to be studied in contact with an atmosphere that is constantly renewed as the gas or oxygen under study is consumed. It consists of a thermostatic chamber in which magnetic stirrers are placed. Hermetically sealed vials are placed on these stirrers, and the sample to be tested is then introduced therein. They have a CO_2 absorber and are connected to an oxygen generator (usually a liquid oxygen cylinder). The depression obtained after oxygen consumption is recorded by a contact barometer. This results in a displacement of the mercury column and the contact sends a pulse that allows the oxygen supply to be turned on or off. This supply is represented in the form of a curve.

6.4.4.1.2. Constant volume respirometers

The Warburg device is the typical apparatus for measuring the respiratory activity of activated sludge. The principle of this apparatus is based on the fact that at a constant volume and temperature, any variation in the quantity of a gas can be measured by the variation of its pressure. A measurement unit consists of a flask, a pressure gauge and various valves connected to the flask and the pressure gauge. During the reaction, the released carbon dioxide is absorbed by a potassium solution contained in a well placed in the flask. Therefore, knowing the volume of gas contained in the flask, the volume of liquid in the flask, the temperature and the density parameters of the manometric liquid, the quantity of gas exchanged can easily be calculated.

Determination of b'

This is the simplest. A known quantity of sludge is introduced into a measuring vial. After heating, the flask is stirred. Given that the sludge in the flask is not fed, only endogenous respiration takes place. The amount of oxygen consumed over time is then directly measured on a pressure gauge. In this case, the coefficient b' corresponds to the slope of the line representing this constant oxygen consumption. Coefficient b' is therefore the ratio of the oxygen consumption established over 24 hours, with the bacterial mass present in the flask.

Determination of a'

In parallel with determining the coefficient b', the coefficient a' can be measured using the same sludge placed in other flasks. To do this, we observe the oxygen consumption curve with respect to time, starting from a vial containing the sludge to be studied and a sample of effluent (for a known quantity). Respiration is therefore accelerated by the presence of organic matter. Once they have been totally degraded, the slope of the oxygen consumption curve thus resumes, after a certain time, with a value identical to that obtained during endogenous respiration, that is, when b' is determined.

Coefficient a' therefore measures the difference between total respiration (oxidation + endogenous) and endogenous respiration during the duration of the test. a' is determined with respect to the amount of BOD in the effluent involved.

Figures 6.5 and 6.6 show the curves obtained when measuring the coefficients a' and b'.

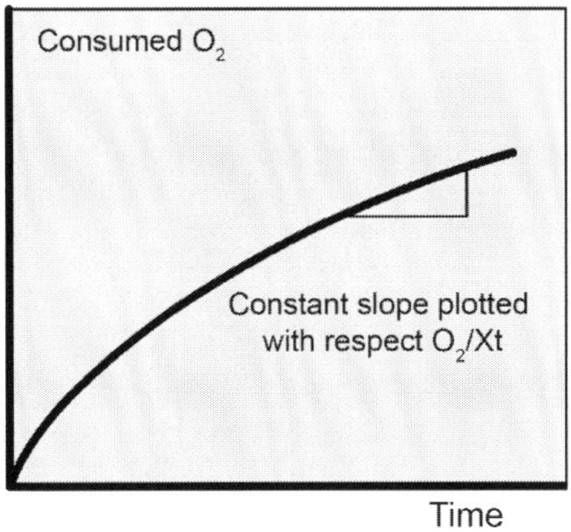

Figure 6.5. *Determination of b'*

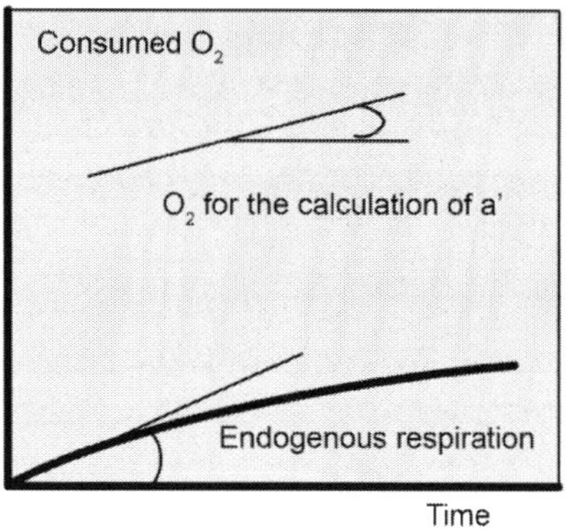

Figure 6.6. *Determination of a'*

Often, for the measurement of a', a series of flasks is used into which samples with different dilutions are introduced in order to properly monitor the oxygen consumption over time. Sometimes a sample that is too concentrated leads to too much oxygen demand, which in this case prevents the respiration curve from being correctly drawn. Several measurement flasks are also used for technical reasons. The measurement of the coefficients a' and b' is time-consuming and requires the constant presence of the manipulator. The use of a series of vials can mitigate deficient tests; for example, a poorly washed flask does not give the same results as a flask that does not have organic residues, detergents or any toxicity.

EXAMPLE.– We assume that the vial contains 2.5 cm^3 of sludge at 4 $g.L^{-1}$ of VSS.

– *Calculation of b'*. It is observed that for 6 hours, the oxygen consumption is 150 μL of O_2. In the equation q O_2/t = a' Le + b' Xt, Le is zero because no feed is provided into the flask. It thus remains: q O_2/t = b' Xt. The total amount of sludge in the flask is: Xt = 4 x 2.5 = 10 mg, that is, Xt = 10,000 μg.

The recorded O_2 consumption over 24 hours is:

$$qO_2 = 150 \text{ x } 24 = 600 \ \mu L.d^{-1}$$

$$qO_2 = 600 \text{ x } 1.43 = 858 \ \mu g.d^{-1}$$

where 1.43 is the density of oxygen under the testing conditions, which is given by:

$$32 \text{ g}/22.4 \text{ L} = 1.43$$

We then have:

$$b' = 858/10{,}000 = 0.0858 \text{ g } O_2.g^{-1} \text{ VSS}.d^{-1}$$

– *Calculation of a'.* After two hours, 1 cm^3 of sufficiently diluted effluent is introduced to provide a BOD_5 of 1,000 µg. There is then a sudden increase in O_2 consumption, and only after 16 hours does the curve become parallel to the endogenous respiration line again. The total O_2 consumption read is equal to 1,020 µL.

The oxygen consumption due to the oxidation of organic matter is:

$$1{,}020 - 600 = 420 \ \mu L \text{ over 24 hours}$$

$$420 \times 1.43 = 600.6 \ \mu g.d^{-1}$$

assuming total BOD_5 removal, that is, Lo-Lf = 1,000 µg.

The value of a' can be assessed as follows:

$$a' = \frac{600.6}{1{,}000} = \frac{0.60gO_2}{g} BOD5.d$$

– *Calculation of b.* During the endogenous respiration phase, we have:

$$C_5H_7NO_2 + 5O_2 \rightarrow 5CO_2 + NH_3 + 2H_2O$$

This reaction shows that 160/113 = 1.42 g O_2/g VSS are required, from which b' = 1.42 b or even b = 1.42 x 0.0858 = 0.12 kg VSS degraded per kg VSS.

6.5. Operating parameters

6.5.1. *Sludge losses: causes and consequences*

Sludge losses can have various origins:

– poor recycling (tank under-design or recycling pump malfunction);

– irregular sludge extraction, which results in an increase in sludge concentration in the settler, making it impossible to separate it. The first losses occur at the time of peak flow.

The operators of the treatment plant are responsible for these sludge losses if they do not regularly monitor the sludge rate and do not regularly perform sludge extractions.

If the flow rate allowed in the treatment units is not limited, then the installation must withstand a strong hydraulic overload during rainfall (unitarian system), and the clarifier, which is conceived to operate at a certain upward flow, will no longer retain the sludge.

The structure of the sludge can also play a role in the clarification. In fact, filamentous organisms may develop, often caused by the nature of the effluent to be treated. Water too rich in sugars or sulfides or too low in nitrogen helps in the development of these organizations. The sludge no longer settles. This phenomenon is called “bulking”.

Sludge treatment facilities can also be incorrectly dimensioned. Indeed, too small a surface area of drying blankets, for example, does not allow a periodic supply of sludge to be dried, thus diminishing the frequency of purges. This then has an impact on the settling.

The same phenomena are also observed when the drying blankets are poorly drained, the aerobic stabilization tank is too small, or the filter material is badly chosen.

The consequences of sludge losses are significant insofar as they directly affect the receiving environment. Sludge discharge is harmful not only to the level of concentration of suspended solids in the receiving environment but also to the level of organic matter, which increases the pollution limit (expressed in BOD or COD). A loss of 55% of the sludge produced by a plant reduces the BOD (or COD) removal efficiency by 50 to 80%.

6.5.1.1. *Measures to implement to avoid sludge loss*

Sludge losses can be avoided by acting at the level of the construction of the facilities, that is, at the design level. The following points should be considered:

– high vertical velocity, which must be calculated at the maximum flow rate that the plant has to treat. For example, in extended aeration, velocities of about 0.8 to 1 $m.h^{-1}$ will be admissible;

– efficient maintenance for recycling pumps;

– a sufficiently efficient settler design: a properly dimensioned sludge pit and a well-designed scraper, capable of bringing the sludge back towards the bottom of the settler;

– a sludge silo could be included, which is always useful during rainy or very cold periods, as it enables best use of the drying blankets by admitting only concentrated sludge and quickly discharging liquid sludge for spreading when required.

6.5.2. *Role of nutrients*

The formula adopted to symbolize living matter can be either $C_5H_7NO_2$ or $C_7H_9NO_3$. These chemical formulas reveal the constituents such as C, H, O and N. Phosphorus should be added, which does not appear in the formula, although various authors have tried to incorporate it. However, the complexity of the equations to be solved has led them to return to the previous formulas.

The overall formula $C_{106}H_{180}O_{45}N_{16}P$ is generally accepted.

It is nevertheless certain that a deficiency in nitrogen or phosphorus in the effluent has a detrimental effect on biological treatment. The absence of nitrogen does not allow bacteria to normally synthesize new cellular materials. When the effluent to be treated is depleted of nutrients (N, P), they should then be added in the form of agricultural fertilizers, either through adding urban effluent in appropriate proportions or in the form of products such as ammonium phosphate. However, this case is the least widespread because of its high production costs.

The action of nitrogen, for example, is clearly obvious when organic compounds such as glucose are removed:

$$21\ C_6H_{12}O_6 + 10\ NH_3 + 56\ O_2 \rightarrow 10\ C_7H_9NO_3 + 56\ CO_2 + 92\ H_2O$$

Generally speaking, the concentration of organic matter in an effluent is expressed in COD or BOD.

Minimum ratios required for biological treatment are then defined:

– in COD: COD/N/P = 150/5/1;

– in BOD: BOD/N/P = 100/5/1.

The urban effluent comprises a surplus of nitrogen and phosphorus and therefore does not require an additive of these elements. We get ratios of about: BOD/N/P = 100/25/10.

6.5.3. *MI*

The MI is the ratio between the volume of the sludge settled in 30 minutes, and the mass of suspended solids contained in this volume:

$$\mathrm{MI} = \frac{\mathrm{V}}{\mathrm{M}}$$

If M is in g, MI represents the volume in cm^3 occupied by 1 g of sludge. It therefore enables us to measure to what extent the sludge is capable of settling. It can also be used to define an interval for the level of SS to be maintained in the aeration tank (Lower X and Upper X) and to calculate the tank volume by bounding it between two values. In practice, the MI can be useful for predicting poor settling, given that it is easy to measure. Although it gives only an approximate value (within 10%) at high concentrations (greater than 5 $g.L^{-1}$), its use remains justified.

6.5.4. *DI*

The DI is the inverse of the MI. It is the density index. Less used than the MI, it nevertheless retains the same justifications:

$$MI.\ DI = 100$$

6.5.5. *Sludge redox potential (Rh)*

When electrons are transferred from one compound to another, an oxidation–reduction (redox) reaction occurs. The donor is the reducing agent, and the acceptor is the oxidizing agent. The redox equilibrium is based on electronic exchanges characterized by an electrode potential calculated with the Nernst equation:

$$E = Eo - \frac{RT}{nF} \text{Log} \left(\frac{Ox}{red}\right)$$

E (in mv) is linked to a hydrogen electrode:

$$Rh = E_H/29 + 2\ pH$$

Variations in the redox potential of the medium can strongly impact the various biological oxidations.

A normally aerated sludge produces an initial potential of about 300 mv. A badly aerated sludge has an Rh of less than 100 mv. In this case, the sludge settles poorly, and the bacterial floc tends to disintegrate. Settling no longer takes place.

The redox potential is measured using specific electrodes. In some cases, redox dyes are used, which change color at a given Rh, thus allowing the detection of local variations in this parameter. The dyes listed in Table 6.15 can be used.

Dye	Rh	E_H at 50% reduction
Toluidine blue	16–18	0.115
Thymol	18–20	0.174
Phenol blue	20–22	0.224

Table 6.15. *Rh and E_H for a few dyes*

A variation in Rh results in a variation in the sludge sedimentation. A phenomenon called "proliferation" then appears, to which we will come back. It is possible to carry out Rh measurements in the field, making it possible to estimate sludge sedimentation.

6.5.6. *Effect of temperature*

The influence of temperature is important in biological phenomena, insofar as it affects the rate of growth of bacteria. The following relation underlines the effect of temperature:

$$K_T = K_{20}\ \alpha^{(T-20)}$$

K_T represents the rate of biodegradation of organic matter at temperature (T°C). It is defined relatively to a reference temperature (20°C or 10°C). The coefficient related to microbial activity (α) considerably varies according to the type of treatment (activated sludge, biofilters, aerated lagoons, etc.). The different values proposed in the literature are summarized in Table 6.16.

This is explained by the effect of oxygen diffusion in the microbial floc at different temperatures. For example, in activated sludge at low loadings and low temperatures, the rate of oxygen use is low, which allows, in a continuous system, greater oxygen diffusion, making the bacterial floc aerobic. When the temperature rises, the rate of use increases due to better assimilation; the oxygen is quickly depleted, making the bacterial floc not aerobic.

Type of treatment	α
Activated sludge: – low and medium loadings – high loading	 1.0 1.0–1.04
Bacterial blankets	1.035
Aerated lagoon	1.06–1.08

Table 6.16. *Coefficient α for various treatment processes*

According to Eckenfelder (2006), with high organic loadings, the floc tends to disperse, and each microorganism is then more directly affected by temperature changes. Coefficient α increases in this case.

The activation energy can also be obtained for two different temperatures, from the equation:

$$K_T = K_{ref}\ EXP\left(-\frac{Ea}{R}\left(\frac{1}{T}-\frac{1}{Tref\text{-}}\right)\right)$$

The activation energy Ea expressed in $Kcal.mole^{-1}$ is deduced from this equation. R is the ideal gas constant, Tref- is the reference temperature taken as 10 °C or 20°C. K_{ref} is the biodegradation rate of the substrate at the reference temperature.

6.5.7. *Effect of pH*

Biological systems tolerate a pH range ranging from 5 to 9 with an optimal zone from 6 to 8.

6.5.8. *Effect of some toxicants*

The toxicants are often of metallic origin. The presence of these toxic substances in the effluent to be treated results in a partial or total inhibition of the microorganism activity. Table 6.17 summarizes some concentrations of metal ions that reduce the effectiveness of activated sludge biological treatment.

Metal	**Metal concentration in raw water (mg. L^{-1})**
Cu	1–2
Ni	1–4
Zn	3–10
Cr (VI)	5–15
Pb	5–15
Hg	0.5–2
Al	2–5

Table 6.17. *Toxic concentration of metals according to bacterial flora*

Providing a concentration level often proves to be challenging, because it depends on the operating conditions of the treatment plant. Different values are obtained when:

– the sludge concentration is variable;

– the toxicant is intermittently provided;

– the substrate is limiting;

– the conditions of the medium (carbonates, pH) are not favorable for these substances to precipitate.

6.5.9. *Sludge settling issues*

Since the MI characterizes sludge by its settling ability, high values indicate poor sludge settling. Different causes are given to this poor settling, which can have various origins.

6.5.9.1. *Dispersed floc*

Activated sludge does not separate from the treated effluent. A cloudy mix that sometimes only clarifies after several hours of rest can be seen. The treated effluent BOD_5 is then very high. This situation can be observed during the period when the microbial flora forms. It becomes serious when this period is too long and leads to sludge deflocculation.

6.5.9.2. *Activated sludge deflocculation*

This phenomenon is sometimes caused under conditions of low applied loading. Actually, the sludge mineralization resulting from the low proportion of organic matter present in the tank, and the long retention times, causes flocs to suspend, which alters the discharge conditions, especially at the SS level, even if, in most cases, the BOD_5 of the treated effluent is satisfactory.

A cause of this phenomenon is the accidental addition of toxic products (mineral or organic) that kill part of the microbial flora. This results in the discharge of an effluent with a high content of suspended solids (inert sludge) and a high BOD_5.

6.5.9.3. *Wave sludge*

Sludge clouds forming in the secondary clarifier can occasionally be observed. These intermittent clouds are often the result of poorly dimensioning the settling tank both at the design and construction level.

6.5.9.4. *Floating sludge*

This phenomenon can be observed both in the test tube and in the facilities. Sludge rises in blocks or fractions that can be seen. There are various causes that explain this phenomenon. The most common is that relating to gas release by denitrification or anaerobic fermentation. At low load, or sometimes even at medium load, part of the ammoniacal nitrogen is transformed into nitrates. When the sludge stays in the clarifier for too long, it uses nitrates as an oxygen source by default. These are reduced to nitrogen gas whose fine bubbles cling to the sludge. Sludge, which is lighter, then rises to the surface. This problem can be overcome by increasing the recirculation rate, unless it is the design of the clarifier itself that is the cause.

In anaerobic fermentation, this process leads to the formation of CO_2 and CH_4. The formed gas bubbles cause the sludge to flow towards the surface and affect the settling.

6.5.9.5. *Foams and bulking*

There are four types of foam:

– Starter or poisoning foams: usually whitish and light, they disappear when the activated sludge reaches maturation or regenerates. They are due to aeration in the presence of partially degraded proteins and detergents. Aeration should be reduced as much as possible.

– Surfactant foams: surfactants come from detergents, colloidal organic matter (blood, for example) or hydrocarbons. If their supply is regular, they can form stable foams. Aeration must be reduced whenever possible and/or use defoamer or spray water to destroy foams.

– Denitrification foams: uncontrolled denitrification during the settling phases results in a decrease in the bulk density of the sludge and its flotation. The resulting floats then form on the surface of the clarifiers and are usually unstable and can be hosed down. This type of foam appears in two specific situations:

- in medium-loading plants when nitrification is possible in summer;
- in low-loading or extended aeration processes where the aeration is in excess.

The solutions consist, for medium-loading plants, in reducing the age of the sludge. For low-loading or extended aeration process, it is recommended to reduce the aeration.

– Stable biological foams: these are stable clumps floating on the surface of the treatment tanks. They are slimy, thick and brown. These foams are not destroyed by agitation. These clusters are flocs formed by the combination of air bubbles and filamentous bacteria whose growth is driven by the presence of greases, easily degradable soluble COD, a substrate deficiency (COD, oxygen, trace elements, vitamins, etc.) or an excessively high sludge age.

6.5.9.6. *Proliferation*

Proliferation is a lightening of the floc due to the appearance of filamentous bacteria such as *Nocardia amarae* (in the event of over-aeration or underload) and *Microthrix parvicella* (if under-aeration or overload). This phenomenon results in poor sludge settling and therefore sludge losses with the treated water.

In summary, there are two kinds of factors driving the development of stable biological foams and bulking: water composition and plant configuration.

The following can reduce bulking and foaming risks:

– the reduction of the level of particulate greases in the influent;

– the reduction of the easily assimilated COD/particulate COD ratio in the inlet of the biological tanks;

– a high COD concentration gradient and a substrate-sufficient medium;

– the increase in sludge settling by ballasting (talc, microsand);

– the design of units without the risk of accumulation of biological foams;

– avoiding the return of fermented sludge to the head of the biological tanks.

Figure 6.7. *Bulking in clarifiers*

Foams and bulking most often appear in low loading and extended aeration tanks or in tanks treating agri-food influents. Particular attention should therefore be given to the design of this type of plant.

The measures generally proposed against the formation of these foams and bulking are given as an indication, bearing in mind that the phenomenon of bacterial growth is not well known.

6.5.9.7. *Redox potential*

It seems that there is a close relationship between the decrease in the redox potential at the floc level and the emergence of bulking. The evolution of this parameter recorded through various chemical tests allows us to interpret, in part, the fluctuations in bacterial populations. It is known that the redox potential is related to the dissolved oxygen content in the medium. An O_2 deficiency creates a biochemical imbalance that is accompanied by a depletion of the medium and impacts the different existing species, thus becoming favorable, for example, to filamentous bacteria.

6.5.9.8. *Composition of the effluent to be treated*

The composition influences the selection of purifying bacteria. This concerns, for example, dairy effluents, in which high concentrations of glucose and lactose often cause sludge proliferation. Lactic acid bacteria that develop in the presence of air, but whose primary energy source is the fermentation of sugars, have an extremely restricted respiratory capacity and cannot produce ATP (adenosine triphosphate) by means of respiration. A reduced oxygen supply for the floc is thereby favorable to them. Germs such as *Lactobacillus* are often found in the proliferation floras.

6.5.9.9. *Nutrient deficiency (nitrogen and phosphorus)*

Nutrient deficiency can cause sludge to proliferate, given the role of these elements in bacterial growth. However, this N or P deficiency is merely one link in a chain and cannot be the only reason for bulking. Other factors can contribute to sludge proliferation, such as temperature (above 30°C or approximately 5°C); pH (acidity contributes to the development of fungi and yeasts); toxic elements (Cr), whose influence is not always harmful; and finally dissolved oxygen, which is linked to the redox potential.

6.6. Activated sludge ecology

The main lines of the classification of living beings enable them to be classified into two categories: the animal kingdom and the plant kingdom. Bacteria are on the border of the two kingdoms. Microflora and microfauna can be identified.

Microflora can account for 10^{11} to 10^{12} of bacteria per g of dry matter. They consist of Gram-negative aerobic mobile bacteria that can travel by way of flagella. Their origin is varied; that is, they originate from the soil, water, etc. We find a variety in activated sludge for the treatment of urban wastewater, such as those represented by Lophomonas, Alcaligenes, Pseudomonas, Arthrobacter, Flavobacter and Achromobacter.

The metabolism of this type of bacteria is the same as that of the other types, namely: pollution adsorption, synthesis and degradation. In these bacteria, it seems that only 15 to 35% of the substrate is catabolized for both sugars and amino acids. Some 20 to 30% is used for cell growth, while the remaining 40 to 60% is used for lipid or polymer secretions with high molecular weight called polysaccharides. Activated sludge appears as a suspension of flaky particles of a few tenths of a millimeter to a few millimeters in diameter (flocs), consisting of bacteria and inert or mineral organic matter maintained by a mucilaginous substance produced by bacterial activity. Normal activated sludge also contains abundant microfauna consisting of protozoa and small metazoans (Sahtout 2012).

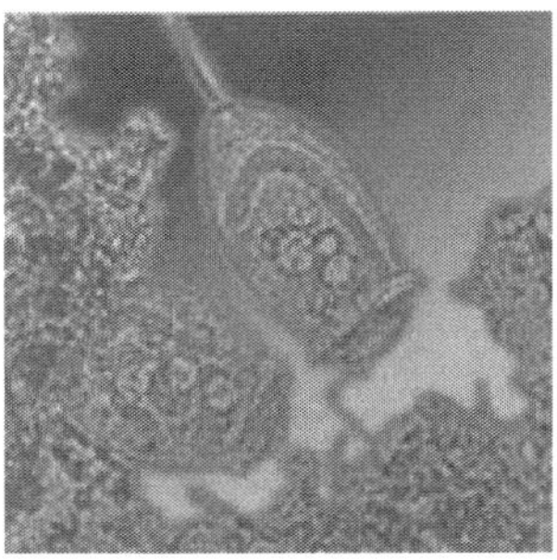

Vorticella

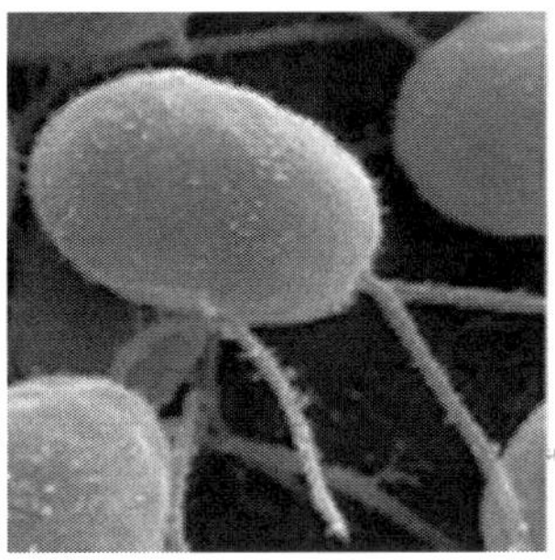

Flagelates

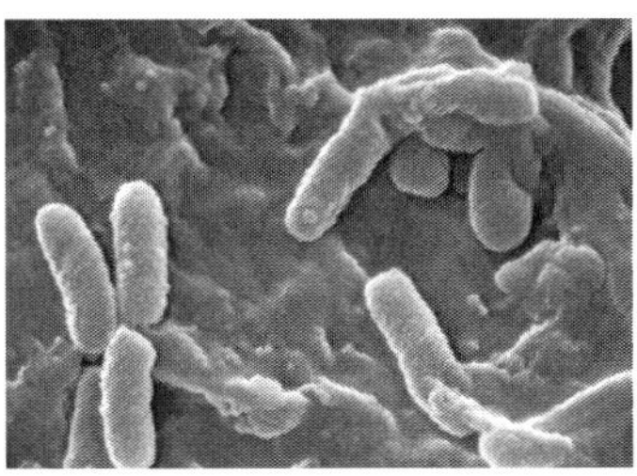

Pseudomonas

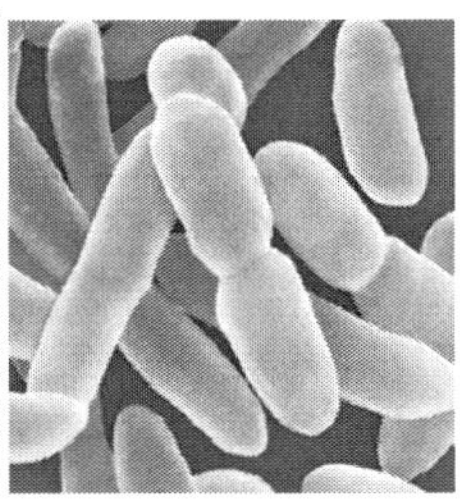

Arthrobacer

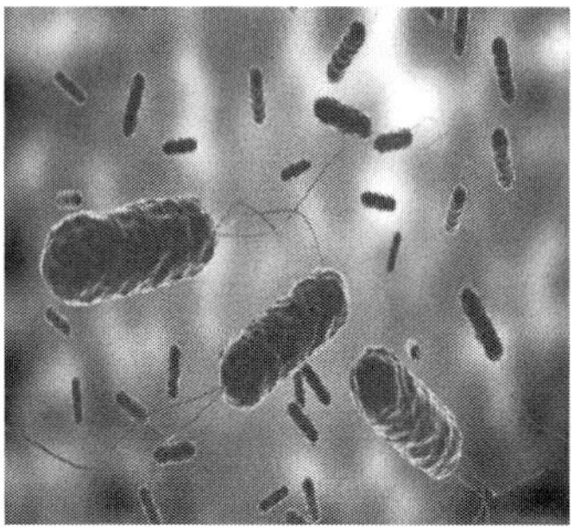

Alcaligenes

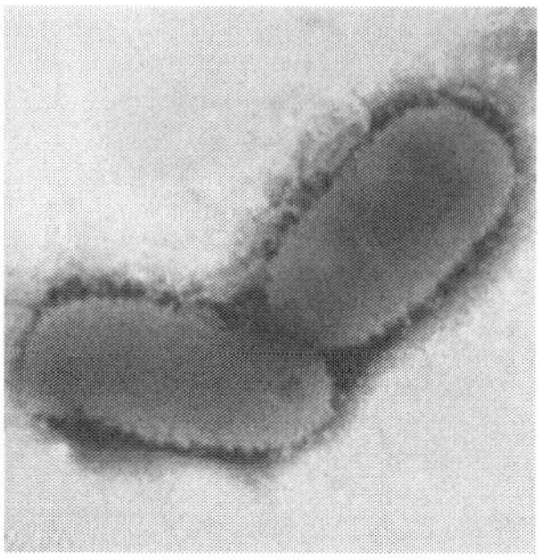

Flavobacter

Figure 6.8. *Microorganisms present in wastewater*

Microbial populations of activated sludge are complex and can only be defined by the main ecological groups: bacteria, fungi, protozoa and metazoans. Most of the purification is due to Gram-negative, mobile bacteria, which include Pseudomonas, Aeromonas, Arthrobacter, Flavobacter, Achromobacter and Alcaligenes. Certain species should also be mentioned, the presence of which can be an issue in that they might be detrimental to sludge settling, namely filamentous bacteria, the most widely known species of which is Sphaerotilus. These bacteria often thrive in nitrogen-deficient media, and they are rich in easily assimilated elements (such as dairy and sugar rejects).

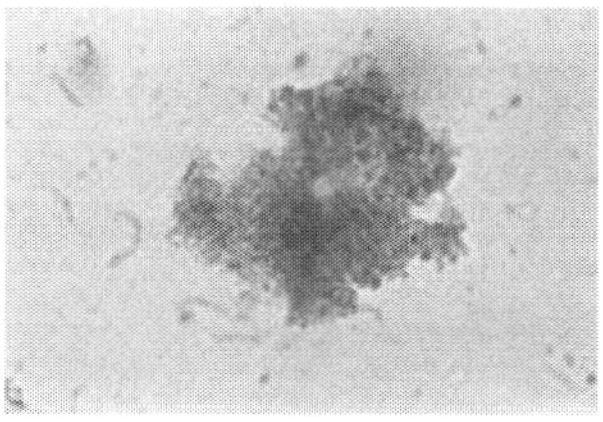

Figure 6.9. *Bacterial floc*

The microfauna of activated sludge are made up of bacteria and microscopic animals, smaller than the millimeter scale. They are mainly represented by protozoa between 20 and 200 μm in size, which include rhizopods or amoebae, zooflagellates, vorticels, opercularia, didinium, lionotus, rotifers, nematodes and parameciums, etc., representing more than 50,000 species. The microfauna of activated sludge can reach a high density in the sludge, namely 10^7 to 10^8 units per liter. The multiplication of these microscopic animals is slower than that of bacteria. In fact, while bacteria divide about every 20 minutes, protozoa divide between two and four times a day. They play a significant role in flocculation, since they eliminate, for example, fecal bacteria, whose particularity is that they do not flocculate.

The nature of the microfauna depends on various factors, such as the chemical composition of the effluent, the proportion of nitrogen presents in the form of a C/N ratio, mass load and retention time.

6.7. Aeration of activated sludge tanks

Active bacteria are essentially aerobic. They therefore use O_2 for substrate assimilation, cell synthesis and endogenous respiration: $C_5H_7NO_2 + 5O_2 \rightarrow 5CO_2 + NH_3 + 2H_2O$.

The ratio of the mass of consumed O_2 to the mass of compounds oxidized by endogenous respiration is about 1.42.

Since the biological treatment is carried out in an oxygenated environment, treatment facilities will include oxygenation devices. In most cases, this is done by transferring O_2 from the air to the liquid phase, which takes place through the interface separating the air from the liquid.

Therefore, in order to degrade organic matter, bacteria consume a certain quantity of oxygen by means of devices called aerators.

It is, however, accepted that aerobic microorganisms suspended in water do not directly use gaseous oxygen, and that the oxygen that is to be supplied to them must first be dissolved in water.

6.7.1. *O_2 transfer*

It can be considered that the phenomenon of oxygen transfer takes place in three distinct stages:

– a first, very quick stage, during which oxygen molecules are transferred from the air to the surface of the liquid;

– a second stage, during which diffusion of oxygen molecules through the liquid film is established;

– a third stage, during which the oxygen is intimately mixed with the liquid by diffusion, convection or even turbulence phenomena: as it happens, for increasing levels of turbulence, the surface of the film is disturbed and the film renewal is responsible for the transfer of oxygen within the liquid.

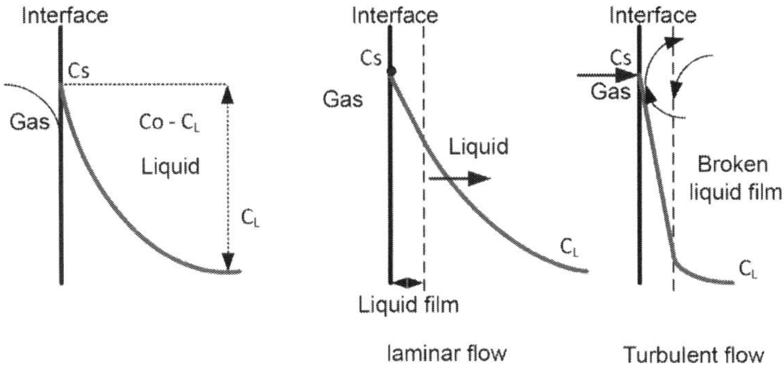

Figure 6.10. *Oxygen transfer mechanism*

Oxygen transfer is governed by Fick's law:

$$\frac{\partial m}{\partial t} = -\mathrm{DL\,A}\,\frac{\partial C}{\partial x}$$

where $\partial m/\partial t$ is the mass transfer velocity; D_L is the mass transfer coefficient (or diffusion coefficient); A is the surface area through which the transfer occurs; and C is the concentration gradient in the liquid layer for a film slice equal to x.

Given that C_L is the concentration of oxygen in the water and Cs is the concentration of saturation oxygen measured at the interface, the equation becomes:

$$\frac{\delta m}{\delta t} = -\mathrm{DL\,A}\frac{\mathrm{Cs} - \mathrm{C}}{1}$$

with $C_L = C$

Writing $K_L = -D_L/l$, where l is the total thickness of the liquid film, we obtain:

$$\frac{\delta m}{\delta t} = \mathrm{KL\,A\,(Cs - C)}$$

By introducing the volume of liquid subject to aeration, it can be written that:

$$\frac{1}{V}\frac{\delta m}{\delta t} = \mathrm{KL\,A}\frac{\mathrm{Cs} - \mathrm{C}}{V}$$

We define a = A/V and K = $K_L a = K_L$ A/V.

A/V indicates that the area A of the interface must be as large as possible with respect to the aerated volume (for instance, fine and numerous bubbles). Multiplying by the interfacial area per unit volume the change in oxygen concentration with time, we then have:

$$\frac{\delta m}{\delta t} = KLa\,(Cs - C)$$

where $K_L a$ is an overall liquid film coefficient, typical of the aerator–tank pair under the testing conditions. It is expressed in h^{-1}.

Writing that the variation in concentration according to time is equal to the variation, per volume unit, of the transfer of oxygen in the liquid as follows:

$$\frac{1}{V}\frac{dm}{dt} = \frac{dC}{dt}$$

the final equation is then:

$$\frac{dC}{dt} = KLa\,(Cs - C)$$

This equation is the basic equation used to describe oxygen transfer in actual aeration systems. The maximum rate of transfer occurs when the dissolved oxygen concentration in solution is very low, close to zero. No transfer occurs when the dissolved oxygen concentration has attained equilibrium with the gas phase. The oxygen transfer coefficient, $K_L a$, is the product of the liquid film coefficient, K_L, and the interfacial area exposed to transfer in a given liquid volume, a.

The saturation value, Cs, is a measured value in aeration systems.

The value of $K_L a$ is used to calculate the quantity of oxygen introduced per unit time (dm/dt), which is therefore proportional to the existing deficit with respect to saturation.

Coefficient $K_L a$ depends on various parameters, such as:

– the volume and the shape of the aeration tank;

– temperature;

– the nature of the liquid to be ventilated.

Both the liquid film coefficient and the interfacial area through which transfer occurs are affected by the type of aeration equipment employed and the turbulence level in the system.

Data available from manufacturers on their specific equipment operating in tap water are typically given as a function of gas flow or power input at a temperature of 20°C. Therefore, the data must be adjusted to account for the temperature in the aeration tank and for the wastewater constituents.

In the previous equation, the parameters that can influence the transfer of oxygen are K_La and Cs. Very often, K_L is simply expressed as K. The equation becomes:

$$\frac{dC}{dt} = K\,(Cs - C)$$

6.7.2. *Influence of some factors on Cs and K_La*

6.7.2.1. *Effect of salt content*

When the water to be treated contains a high salt content, the value of C_S is obtained with the relation:

$$Cs\ (g.L^{-1}) = \frac{475 - 2.65\ S}{475}$$

where S is the salinity of the water ($g.L^{-1}$).

6.7.2.2. *Effect of detergents and oils*

The value of Cs is reduced in the presence of detergents, oil, greases or suspended solids. This parameter is corrected by a multiplying coefficient of about 0.95 to 0.99.

6.7.2.3. *Effect of temperature*

Cs decreases as the temperature increases. The proposed formula to correct this effect is:

$$Cs\ (mg.L^{-1}) = \frac{475}{475 + T}$$

Temperature T is expressed in °C.

6.7.2.4. *Gas solubility in liquids*

Gas solubility in fresh water mainly depends on temperature, the type of gas and the partial pressure of the gas. In addition to these three parameters, gas solubility in water also depends on salinity. According to Henry's law, the amount of air dissolved in saturated water depends on the temperature and pressure of the air above the liquid. At a constant temperature, the amount of gas dissolved in a liquid is proportional to the pressure exerted by that gas on the liquid. With Henry's law, the fraction of a mole of dissolved gas of any gas, xi, at equilibrium conditions can be calculated as:

$$xi = Pi\ (h)/Hi$$

with:

– xi being the molar fraction of air dissolved in water ($mole.mole^{-1}$);

– Pi (atm) being the partial pressure of the gas phase above the liquid (in this case water) at an elevation h (h = 1.0 if the elevation is 1,000 m);

– Hi (atm) being the Henry's law constant.

It is linked to the nature of the dissolved gas (air here). It depends on the temperature T. It increases as the temperature increases. The mole fraction of dissolved air decreases as the temperature increases:

$$H_T = H_{10.}\ 1.02^{(T - 10)}$$

where T is expressed in °C.

The partial gas pressure is corrected according to the altitude:

$$Pi(h) = Pi(o)\ e^{-0.121h}$$

with:

– h being the altitude above sea level in thousands of meters;

– Pi(o) being the partial gas pressure, i, at sea level where h is 0 m. The partial pressure of O_2 in the atmosphere is 0.2095 atm.

The concentration of dissolved gas ($mole.L^{-1}$) is calculated using the following approximation:

$$Ci = Xi\ Cwater$$

with:

– Ci being the dissolved gas concentration in mole. L^{-1};

– Cwater being the molar concentration of water in mole. L^{-1}; Cwater = 1,000 ($g.L^{-1}$) / 18 ($g.mole^{-1}$) = 55.56 $mole.L^{-1}$.

For example, the molar fraction of O_2 in water at atmospheric pressure and at 20°C is:

$$X_{O2} = P_{O2}/H_{O2}$$

That is:

$$X_{O2} = 0.2095\ (atm)/(4.09.\ 104)\ (atm) = 5.12.10^{-6}$$

The saturation concentration of dissolved O_2 (20°C) is therefore:

$$C_{O2} = 5.12\ 10^{-6}\ x\ 55.56\ x\ 32{,}000\ (mg.L^{-1}) = 9.11\ mg.L^{-1}$$

The dissolved air concentration is the sum of the respective saturation concentrations of O_2 and N_2. That is:

$$Cs,air = Cs,O_2 + Cs,N$$

The saturation concentration of N_2 is calculated using the same method as that of O_2. The molar fraction of N_2 in water at atmospheric pressure and at 20°C is:

$$X_{N2} = P_{N2}/H_{N2}$$

That is:

$$X_{N2} = 0.7809\ (atm)/(8.04\ 10^4)\ (atm) = 9.71\ 10^{-6}$$

The saturation concentration of dissolved N_2 (20°C) is thus:

$$C_{N2} = 9.71\ 10^{-6}\ x\ 55.56\ x\ 28{,}000\ (mg.L^{-1}) = 15.1\ mg.L^{-1}$$

The dissolved air concentration at 20°C is thus:

$$Cs,air = 9.1 + 15.1 = 24.2\ mg.L^{-1}$$

Table 6.18 gives some values of $H.10^4$ (atm) for dissolved gases in water.

Temperature (°C)	Air	CO_2	H_2S	N_2	0_2
0	4.32	0.073	0.027	5.29	2.55
10	5.49	0.104	0.037	6.68	3.27
20	6.64	0.142	0.048	8.04	4.09
30	7.71	0.186	0.061	9.24	4.75

Table 6.18. *Henry's constant for dissolved gases according to temperature*

Temperature (°C)	Dissolved oxygen (mg.L-1)	Temperature (°C)	Dissolved oxygen (mg.L-1)
0	14.60	23	8.56
1	14.19	24	8.40
2	13.81	25	8.24
3	13.44	26	8.09
4	13.09	27	7.95
5	12.75	28	7.81
6	12.43	29	7.67
7	12.12	30	7.54
8	11.83	31	7.41
9	11.55	32	7.28
10	11.27	33	7.16
11	11.01	34	7.05
12	10.76	35	6.93
13	10.52	36	6.82
14	10.29	37	6.71
15	10.07	38	6.61
16	9.85	39	6.51
17	9.65	40	6.41
18	9.45	41	6.31
19	9.26	42	6.22
20	9.07	43	6.13
21	8.90	44	6.04
22	8.72	45	5.95

Table 6.19. *Solubilization of dissolved O_2 according to temperature (below 760 mm Hg)*

The Henry constant can also be deduced from the relation:

$$Ci = Pi / Hi$$

or:

$$Hi = Pi / Ci$$

with C expressed in $mole.L^{-1}$ and P in atm. In this case, the unit of Hi is $L.atm.mole^{-1}$.

As the temperature increases, the oxygen saturation concentration in the water decreases.

6.7.2.5. *Effect of atmospheric pressure*

Under an atmospheric pressure (Pa) in water (m), the correction to be made is P/10.33. For different altitudes, Table 6.20 shows the corrections to be made.

Altitude (m)	0	500	1,000	1,500	2,000	2,500	3,000	3,500	4,000
Pa (mm Hg)	760	716	674	630	596	560	526	493	462
m water	10.33	9.74	9.17	8.63	8.10	7.63	7.15	6.71	6.20

Table 6.20. *Pressure according to altitude*

By taking these various parameters into account, it is possible to calculate the corrected value of Cs.

6.7.2.6. *Influence of some factors on KLa*

As for Cs, various external parameters influence the value of K_La. These include the salt content of the water to be treated, the temperature, the atmospheric pressure or even organic or mineral products such as detergents, greases and suspended solids.

– The influence of temperature shows that the K_La coefficient increases when there is an increase in T °C. Temperature increases K_La similarly to the effect on diffusivity and the liquid film coefficient, as indicated by the following relationship.

θ is dimensionally not homogeneous, requiring a temperature in °C. An alternative would be to express the temperature impact in the exponential form.

The relation between K_La and T °C is as follows:

$$K_La_{T''} = K_La_T\ \theta^{(T'-T)}$$

where K_La_T is the oxygen mass transfer coefficient at T, (s^{-1}); θ is a temperature coefficient; T' is the working temperature (°C); and T is the reference temperature (°C). It is usually taken as 20°C.

That is, $K_La_{T'} = K_La_{20}\ 1.02^{(T'-20)}$

K_La_{20} is the oxygen mass transfer coefficient at 20°C (s^{-1}).

Depending on the manufacturer, the coefficient θ can vary between 1.047 and 1.016.

– The influence of the water quality also affects the value of K_La. Since temperature also affects viscosity and surface tension, changes in the interfacial area as well as K_La may also result. It can be shown that they concentrate on the liquid surface, preventing the diffusion of oxygen into the water. The transfer coefficient of polluted water is therefore lower than that of clean water.

We define a coefficient σ to account for the wastewater effects. σ is the ratio between the value of K_La in wastewater and that of K_La in clean water, such that:

$$\sigma = K_La\ \text{wastewater} / K_La\ \text{clean water}$$

At equilibrium, the excess surfactants on the liquid surface can be defined by Gibbs's law. This equation shows that there is a maximum concentration on the surface of the liquid:

$$\beta = \frac{C}{RT}\frac{dY}{dC} = Cs_{20}(\text{wastewater}) / Cs_{20}(\text{clean water})$$

where:

– $C_{S,20}$ (wastewater) is the saturation with oxygen dissolved in wastewater at 20°C, determined in an aerated mixture of wastewater and sludge (g O_2/m^3);

– $C_{S,20}$ (clean water) is the saturation with oxygen dissolved in water at 20°C, determined in an aerated clean water (g O_2/m^3).

Water rich in surfactants will tend to decrease the value of σ. This coefficient therefore depends on the nature of the water, which itself affects the diameter of the air bubbles in biological tanks. The numerical values of σ vary between 0.2 and 0.9 depending on the content of detergents, defoamers and albuminoid materials in urban wastewater.

	Large bubbles Surface aerator	Medium bubbles	Fine bubbles
Activated sludge	0.9	0.8	0.7
Fresh sludge	0.6	0.5	0.4

Table 6.21. *Coefficient σ according to the type of aeration*

This is explained by the fact that the presence of surfactants causes a diminution in the diameter of the bubbles, which corresponds to an increase in the ratio of the transfer area/liquid volume. Sometimes the value of σ is greater than 1 for high surfactant concentrations.

EXAMPLE.– After water aeration in a tank, the decrease in oxygen is monitored as a function of time. The aeration of deoxygenated water gives the values, presented in Table 6.22, of the oxygen deficit according to time.

Time (min)	O_2 deficiency (C_S-C_L) ($mg.L^{-1}$)
0	10.6
6	8.4
12	7.2
18	6.2
24	5.3
30	4.6
36	4.0
48	2.9
54	2.5
60	2.2

Table 6.22. *O_2 deficit results versus reaction time*

The oxygen transfer equation is:

$$\frac{dC}{dt} = KLa\ (Cs - CL)$$

C_L = C to time t

$$KLa = \frac{Log(Cs - CL)1 - Log\ (Cs - CL)2}{t2 - t1}$$

By plotting Log (Cs – CL) with respect to time, we can easily determine K_La by the slope of the line, that is, $K_La = 1.62\ h^{-1}$.

EXAMPLE.– What is the mass transfer coefficient at 30°C if $K_La_{20} = 1.4\ h^{-1}$?

$$K_La_{30} = K_La_{20}\ 1.02^{(30-20)}$$

That is:

$$K_La_{30} = 1.71\ h^{-1}$$

The biological treatment also affects α since it shifts from 0.72 to 0.90 after four hours of aeration with an urban effluent.

Oxygen transfer in water depends on the performance of the aerators and therefore of the aeration system. Air contains 23% by weight of oxygen, or 20.8% by volume. The specific weight of air (ρa) can be calculated from:

$$\rho a = P/RT$$

where ρa is the specific weight of the air ($kg.m^{-3}$), P is the absolute pressure (mm CE), T is the absolute temperature (degrees Kelvin), and finally R is the specific constant of the air (29.27).

At 0°C, and with a mercury column of 760 mm, 1 m^3 of dry air (ρa = 1.293 $kg.m^{-3}$) contains 300 g of O_2. It contains only 280 g at 20°C (ρa = 1.206 $kg.m^{-3}$).

The introduced oxygen values expressed as a percentage of the oxygen volume are therefore not comparable with aeration under pressure, unless the reference pressure and temperature are known.

It is therefore important to know the aeration system, that is, the way in which air is introduced and distributed in the water, as well as the way in which air bubbles are kept in the liquid.

6.7.3. *Aeration systems*

The differentiation between the treatment systems is mainly the result of the ways that the air is introduced. They include:

– pressurized air aerators blowing air into the water at different tank depths. The insufflation is carried out by means of compressors;

– mechanical surface aerators creating a turbulence zone on the surface of the water. The air is injected at the level of the turbulence zone;

– turbine aerators in which air is blown from a tubing towards the rotating blades of a turbine. The air is dispersed throughout the tank;

– compressed air aerators.

Aerators must ensure minimum turbulence in order to prevent sludge deposits. To this end, a minimal power must be supplied to the terminals, of about 30–40 W per m^3 of tank.

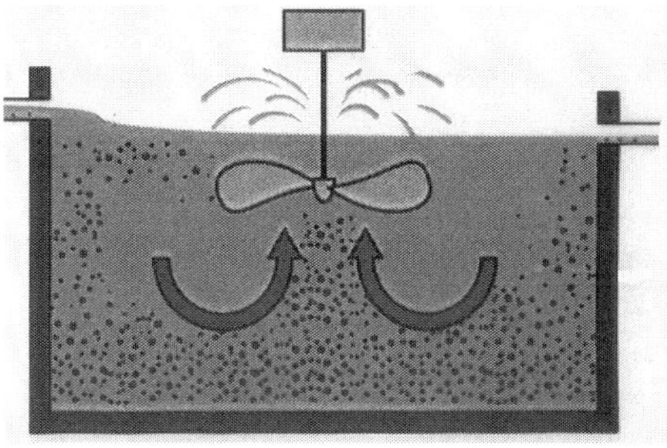

a)

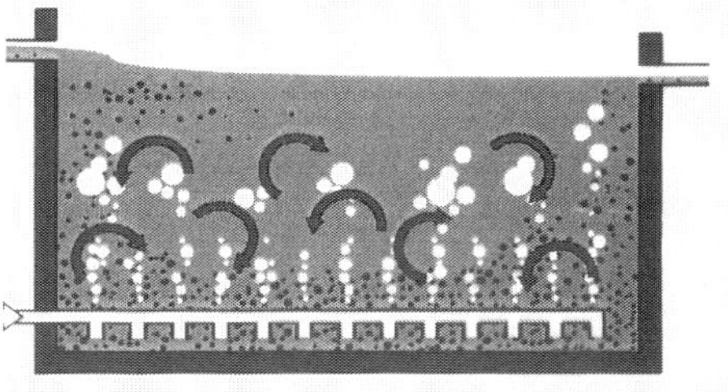

b)

Figure 6.11. *a) Surface aerator. b) Air diffusers*

6.7.3.1. *Over-compressed air aerators*

These are also called air insufflators and are supplied by blowers.

There are three types of compressed air or air diffusion aerators: fine, medium and large bubble diffusers.

These diffusers consist of porous tubes made of aluminum oxides, an equivalent material or a membrane material.

Optimal aeration is obtained by using fine bubbles diffused by flexible membranes such as oxytubes. The resuscitators are generally placed in 4- to 9-meter-deep tanks. In order to minimize the risk of clogging and aging too quickly, the temperature, flow, pressure and cleanliness of the air supplied should be monitored early from the design stage, followed by the necessary maintenance that needs to be carried out.

It is imperative to install a purge system at the air duct level.

Pressurized air aeration consists of blowing air into the liquid mass at depths of up to 7 m in some cases. The systems used are divided into large families according to the size of the generated bubbles:

– large bubbles ($d > 6$ mm): vertical rods, diffusers with large pores;

– medium bubbles ($4 \text{ mm} < d < 6$ mm): various diffusers that reduce the size of the released bubbles (valves, small pores, etc.);

– fine bubbles ($d < 4$ mm): air diffusion through porous bodies or finely perforated elastic membranes. This is called floor design or layout.

Large bubble diffusers are designed with large pores such that large air bubbles can escape. Regardless of the type, diffusers are distributed over the surface of the aeration tank. They can also be installed over a strip allowing a gyrating movement, known as spiral flow. Medium bubbles have a diameter of approximately 3 mm, whereas large bubbles often exceed 6 mm. The advantage of large bubble aerators is that they do not clog; they maintain the temperature of the liquid and incur low maintenance costs. The oxygen transfer rate, however, remains low.

Fine bubble aerators have the advantage of being highly flexible and having a good capacity for mixing and maintain temperature. Their main disadvantage is their high purchase and maintenance costs, because they require filters and auxiliary equipment to avoid clogging.

The bubble diameter obtained by diffusion depends on the power of the jet and the diameter of the holes. The diameter d_B of the bubbles produced by a diffuser is related to the gas flow rate by the relation:

$$d_B = G^n$$

where n is a coefficient related to the structure of the porous material.

The number of bubbles generated is equal to the flow rate divided by the volume of the bubbles:

$$6G/\pi d^3_B$$

The area of bubbles generated is equal to this number multiplied by the area of each bubble:

$$\frac{6G}{\pi dB^3}\pi d^2B = \frac{6\ G}{dB}$$

Other physical parameters can influence bubble diameter. Large bubbles can be categorized into fine bubbles, while fine bubbles conversely can form large bubbles. This is the phenomenon of coalescence. This is avoided by arranging the diffusers according to minimum and maximum spaces to, on the one hand, keep the solids in suspension and, on the other hand, prevent the coalition of bubbles while maintaining the necessary aeration.

The retention time of a bubble in the aerator is: H/v_B, where H is the immersion height of the aerator and v is the ascent rate of the bubble.

The liquid/gas interface area is given by:

$$A = \frac{6G}{dB}\frac{H}{vB} = 6\ G^{1-n}\ \frac{H}{vB}$$

If we call L the tank length, the volume per unit length is L x H, which leads to the relation:

$$\frac{dmB}{dt} = BG^{1-n}\ \frac{H^{1-p}}{L^q}\ (Cs - C)$$

The values of B, p, n and q are experimentally obtained. N varies between 0.5 and 1, p and q are close to 1 and B is characteristic of the bubble system. A good proportion can be obtained, for example, by adopting a tank width twice the depth. Two lateral strips or a central strip of diffusers are then provided.

Aerator	Oxygen transfer efficiency ($gO_2.Nm^{-3}.m^{-1}$)	O_2 efficiency (kg $O_2.kWh^{-1}$)
Fine bubbles	9-15	1.5-3.0
Medium bubbles	5-9	1.0-1.8
Large bubbles	4-6	0.7-1.2

Table 6.23. *Specific uptake and aerator efficiency*

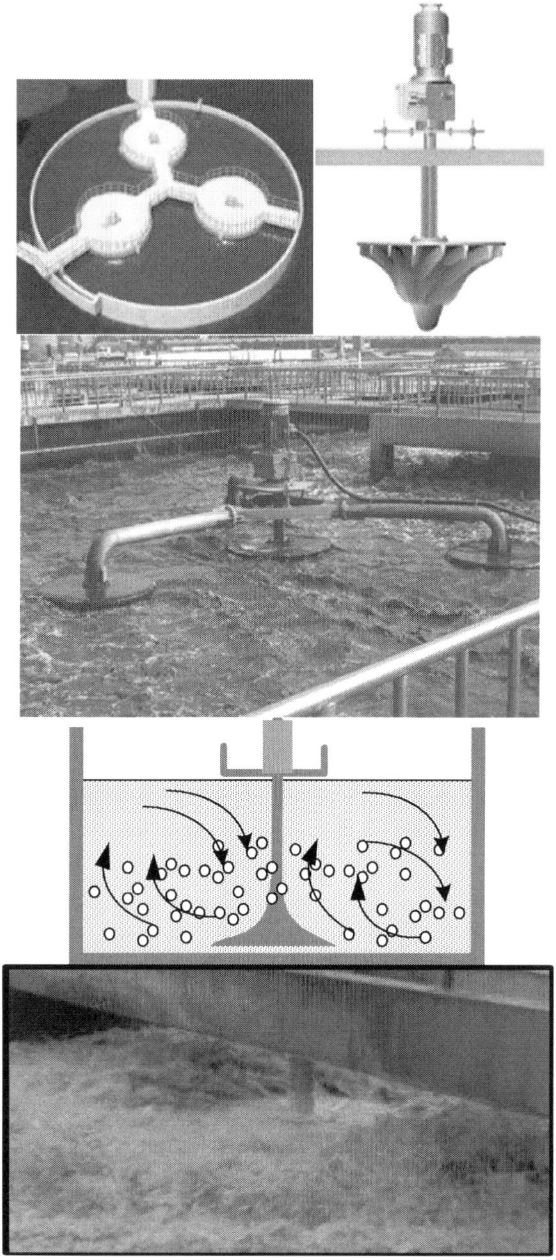

Figure 6.12. *Agitation aeration*

Diffuser efficiency depends on the system adopted: with large-bubble devices, the oxygenation efficiency is approximately 6 to 7%, while that of medium bubbles varies from 6 to 7% in pure water and from 5 to 6.5% in the presence of sludge. Finally, the efficiency of fine bubbles is 20% in pure water and 15% with sludge. In fact, the efficiency depends on the shape of the tanks, with an advantage for traditional (rectangular) tanks compared to combined tanks with an inclined slope. For example, estimates relative to oxygen supply in pure water bring forward the advantage of fine bubbles.

Figure 6.13. *Diffuser disposition in an aeration tank*

Eckenfelder proposes the following equation to define the performance of air diffusers (Eckenfelder 1959):

$$N = C\,Gs^{1-n}\frac{Hm}{Wp}\,(Cs - C)1.02\,\alpha^{T-20}$$

where N is the oxygenation capacity (OC; $gO_2.m^{-3}.h^{-1}$), Gs is the air flow rate, W is the tank width (m), H is the tank depth (m), Cs is the dissolved oxygen concentration at saturation ($mg.L^{-1}$), C is the concentration of dissolved oxygen in the liquid ($mg.L^{-1}$), T is the temperature (°C), α is the oxygen transfer coefficient in the medium under consideration, and n, m and p are the characteristic constants of aerators.

Fine bubble diffusers, for example, are blamed for frequent clogging caused by the air or simply by activated sludge. Generally speaking, diffuser maintenance is complex and requires excessively long times for dismantling and repair. Currently, the diffusers are embedded in trays that can be raised to the surface using elevators and which avoid emptying the tank.

6.7.4. *Surface aerators*

These are classified into two main categories:

– horizontal-axis turbines (also called brush aerators);

– vertical-axis turbines with peripheral blades.

This equipment is effective for tanks with a depth of less than 5.5 meters, this value varying according to the characteristics and size of aerators as well as the conditions of implementation.

The immersion depth of surface aerators greatly influences the amount of oxygen introduced. Variations in immersion depth must therefore either be taken into account for the control of aeration or minimized by designing weirs capable of limiting variations in water level in the tanks.

Surface aerators are preferably used in clocked operations with limitation of the variation in the water body in order to maintain an optimal depth of immersion.

The water level control by mobile weirs proposed by Krüger enables using surface aerators in continuous operation by taking advantage of the variation in the permissible depth of immersion of the aerators.

Independently of the type of turbine, the oxygen supply is carried out in the turbulence zone created by the blades. The turbulence phenomenon permanently renews the interfaces between the air bubbles and the water, thus resulting in intense oxygenation.

The rotating brush, usually made of steel, sprays the liquid using rotary plates, creating a significant gyratory motion. As a result, especially for brushes, the shape of the tank must be designed in such a way that it can ensure the gyration and rotation of the brushes. The rotation of the brushes causes the effluent to move, introduces air bubbles into the liquid phase and creates surface turbulence. The oxygen transfer efficiency (OTE) varies between 1.3 and 1.9. The diameter of the rollers varies from 0.7 to 0.9 m, and the volume power dissipated is approximately 30 $W.m^{-1}$. With 0.7 m diameter rollers, the peripheral velocity at the tip of the blades is about 4 $m.s^{-1}$. The advantages of these systems lie in their good mixing capacity and good oxygen transfer rate, but their purchase cost remains high.

Aerators can be installed on walkways overlooking the water body of the aeration tank or can be supported by floating equipment. The latter technique is suitable for large ponds.

The performance of surface aerators will depend on various factors such as velocity and diameter of the rotating elements, the air flow rate and also the depth of immersion of the equipment.

Turbines can be high-speed (1,500 rpm^{-1}) and vertical axis turbines. They are economical and easy to maintain but have poor characteristics (OTE: 0.8–1.2). These aerators are directly driven by an electric motor at 750 or 1,500 rpm^{-1} without intermediary reduction gear. A small-diameter propeller is usually placed inside a short chimney. The mechanical assembly is frequently supported by one or more floats.

High-speed and horizontal-axis turbines comprise a propeller located underwater (and not at the air–water interface) and attached to a solid or hollow shaft. The other end of the shaft is out of the water and attached to the electric motor. When the propeller is spinning at high speed, between 1,800 and 3,600 rpm^{-1}, a depression forms around it. In this way, the air is sucked in, mixed with the water and discharged. The advantages of these systems are the absence of water projection and limited tank sizes, low purchase costs and good mixing capacity. In addition, they are portable and generate less noise than other systems. The disadvantages are the poor ability to transfer oxygen and issues associated with the mechanical design.

Figure 6.14. *Brush aerators*

The two advantages of these systems are their moderate price and great flexibility of use. They are, however, energy-intensive and have low stirring capacity, and oxygen transfer is low. They are better suited for lagoon than for activated sludge tanks, where deposits should be avoided.

Concerning slow turbines (50–200 $tr.min^{-1}$), performance is significantly improved whether in open or closed systems. The oxygen efficiency of these turbines varies between 1.5 and 2 kg $O_2.kWh^{-1}$. Some installations may transfer more than 2 kg $O_2.kWh^{-1}$; this depends on the shape of the tank, the surface of the composition of the wastewater and the operating conditions.

The role of surface aerators is twofold, since they must provide oxygen and also ensure mixing; in other words, they should keep the sludge in suspension in the tank.

It is important to keep in mind that the tank depth is to be respected, since beyond a given value, mixing is no longer ensured. In this case, part of the sludge settles at the bottom of the tank, creating dead zones, which can have disastrous consequences for the treatment. Maximum depths of 5 m are allowed. They depend on the type of system implemented.

Figure 6.15. *Slow-speed turbine*

6.7.5. *Deep-water aerators*

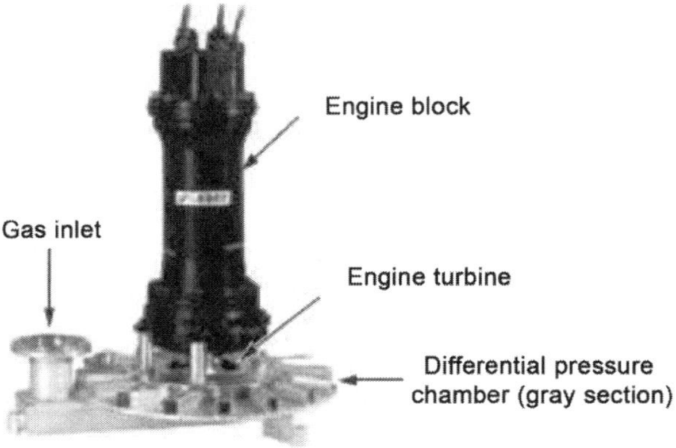

Figure 6.16. *Deep-water aerator (TA of ABS)*

These aerators, placed at the bottom of the tanks (Figure 6.16), generally consist of:

– an electrical submersible motor block;

– a circular pressure-reducing chamber consisting of a double plate connected to an air suction tube. The upper part of this chamber is equipped with a multitude of horizontal diffusers radially arranged on the plate;

– a turbine for the pumping of water, the supplying of air and the discharge of the two-phase mixture obtained.

The rotation of the turbine inside the chamber generates a vacuum in the lower part, which causes a significant quantity of air to be introduced through a suction pipe connected to the outside. At the same time, the hollowed-out vanes of the turbine ensure the pumping of the liquid through its upper part. Therefore, an intimate mixture is obtained by shearing the air into fine bubbles. The discharge of this two-phase mixture is radially carried out through the diffusers placed horizontally.

6.7.6. *Conditions of use of aeration systems*

Particular attention should be paid to the conditions of use and the technical difficulties found in wastewater treatment plants. In general, maintenance is limited to the usual maintenance checks of all engines such as oil level and bearing lubrication. This is not enough because the aerators constantly generate shocks and vibrations to the bearings and gears of the drive unit during operation. In the long term, this induces an imbalance in the system and impacts the efficiency of the aerator and its mechanical system. Moreover, the location of the aerators must be studied in such a way that its distance from the edge of the tank is optimal. If it is too large, the oxygenation efficiency decreases because the air entering the water through the turbulence zone tends to escape as the flow section towards the wall increases.

Table 6.24 shows the minimal powers to be used to prevent settling.

Type	Minimal specific power ($W.m^{-3}$)	Maximal depth (m)
Turbines	30	3.3
Brushes	25	2.2
Insufflation of fine bubbles	12-15	> 3.5

Table 6.24. *Minimal powers to be implemented*

Energy savings are achieved when good control of the oxygenation is implemented in accordance with the requirements necessary for the degradation of organic materials.

To maintain the plant as clean as possible, splashing over the walkways and supports must be avoided by providing a sufficient distance between the water surface and the aerator.

The conclusions reached on the use of aerators are mainly based on the fact that there are technical conditions for the aerator and the tank that must be complied with.

Three essential parameters can be distinguished:

– Immersion: each device requires optimal immersion that corresponds to its best performance. Knowing the quantity of oxygen to be provided, the immersion of the aerator should be adjusted to vary the oxygenation conditions.

– Tank volume: the best oxygenation conditions are obtained for small tank volumes. Volume plays its role insofar as the OC of an aerator is linked to its specific power (power absorbed per m^3 of tank).

– Tank shape: the operating mode of the aerator requires a suitable shape for the tank. Brushes require elongated tanks (with baffles), while vertical-axis aerators are more efficient in compact tanks, with a depth equal to about half the length (square shape, for example).

Aeration mode	Tank depth (m)	Oxygen uptake ($g.kWh^{-1}$)
Porous plates	4	1,800-2,000
Perforated tubes	4	1,000
Simplex turbine (I)	5.5	1,800
Simplex turbine (II)	1.8	2,000
Kessener brush (I)	2.9	1,800
Kessener brush (II)	1.0	1,300
Blade rotors	2.1	1,300
Sheffield vane rotor	1.1	1,800

Table 6.25. *Characteristics of some aeration equipment*

An aerator is chosen according to various criteria such as:

– high specific oxygen uptake (kg O_2/kwh);

– intense mixing of the water mass;

– complete homogenization of the activated sludge and wastewater;

– insensitivity to corrosion;

– horizontal projections at the aerator outlet;

– adjustments of the OC according to needs;

– quick disassembly and easy mechanical operation.

6.7.6.1. *Aeration methods*

Oxygen requirements for activated sludge are calculated for the daily nominal pollution flux and for the dry-weather hourly peak flow. Aeration devices must be able to meet peak oxygen requirements without over-aeration during periods of minimum requirements. Aeration in a tank can be permanent or sequenced.

6.7.6.1.1. Permanent aeration

Permanent aeration helps maintain aerobic conditions in the tank. It is suitable for use in tanks where only carbonated pollution needs to be removed or nitrification is also to be achieved (denitrification is then carried out in a separate non-aerated tank).

The air supply can be done:

– Continuously: the aerators operate at a variable flow rate according to a guideline of dissolved oxygen concentration in the tanks.

– Cyclically: the dissolved oxygen concentration must vary between two thresholds above zero. This mode of operation can be set up with surface aerators (common operation mode) or insufflators. Installing systems that vary the supplied air flow is therefore not essential. The installed aeration system must be designed to withstand frequent restarting. Particular attention must be paid to the tightness of the insufflators (no porous diffusers) and to the number of starts allowed per hour for the blowers. The cyclic mode of operation prevents over-aeration.

6.7.6.1.2. Sequenced aeration

Sequenced aeration is discontinuous aeration that allows the nitrification and denitrification phases to be linked together in the same tank. In sequenced aeration carried out by insufflation, the flow of air blown does not have to be varied. As with cyclic operation, the control must be adapted to the permissible restart frequency of the equipment (booster, turbines or brushes motors). Particular attention must also be paid to the tightness of the insufflators (no porous diffusers and regular maintenance of membrane diffusers). Permanent aeration with cyclical operation should not be mistaken for sequenced aeration. In the first case, aerobic conditions are maintained in the tank; only the mode of operation of the aerators is cyclic. In the second case, aerobic and anoxic phases are performed in the same tank.

6.7.6.2. *Performance of aerator-tank pairs*

6.7.6.2.1. Characteristics of aerators under standard conditions

The performance of the aerators should be defined under standard conditions to facilitate the comparison of their performance.

The standard conditions are as follows:

– clean or clear water;

– initial dissolved oxygen concentration = 0 mg.L^{-1};

– water temperature = 10–20°C;

– atmospheric pressure = 760 mm Hg.

6.7.6.2.2. Transfer factor (TF)

The performance expected from an aeration system under real conditions can be deduced from that determined under standard conditions by means of the transfer factor (TF), which takes into account the differences between the two states:

– mixed liquor instead of clear water;

– temperature other than 20°C;

– atmospheric pressure other than 760 mm Hg;

– dissolved oxygen concentration in the aerated tank other than 0 mg.L^{-1}.

It is used to calculate the standard oxygen requirements according to the actual needs:

Standard oxygen requirements = actual requirements/TF

The TF is influenced by:

– the bubble diameter: in fact, the decrease in the diameter of the blown bubbles leads to an increase in the exchange surface and therefore in the standard insufflator aeration stabilization basin (ASB) and also a decrease in the factor α and therefore in the TF factor. Consequently, it is therefore necessary to study what effect will be predominant in each case:

– water salinity, which slightly decreases the value of β;

– water temperature;

– the presence of surfactants and detergents: these products tend to form air bubbles that coalesce less but are isolated from the water by a thin film. The effect on oxygen transfer may be different depending on the case.

The TF is not influenced by the pH when it is between 5 and 9.

6.7.7. *Aerator design*

The installed aeration system must meet air requirements:

– in nominal operations;

– in maximum operations during the hourly flow and pollution peaks;

– in minimal operations (under-loaded plant and minimum pollution flow).

These requirements can be met by continuous or discontinuous operation of the aerators.

6.7.7.1. *Calculation of the power to be installed or the flow rates to be supplied*

6.7.7.1.1. Surface aerators

The power to be installed for surface aerators is determined from the actual oxygen requirement, the TF and the standard ASB as indicated here:

– determine the standard oxygen mass requirement:

standard O_2 requirements = actual requirement/ TF

– calculate the power of the aerators to be installed:

power to be installed = standard O_2 requirements/standard ASB

power to be installed = real O_2 requirements/(TF * standard ASB)

with:

- O_2 requirements in $kgO_2.h^{-1}$;
- standard ASB in $kgO_2.kWh^{-1}$;
- dimensionless TF.

6.7.7.1.2. Air insufflators

The air flow rate to be blown depends on the (standard) oxygenation efficiency in clear water as shown here:

– determine the standard oxygen mass requirement:

standard O_2 requirements = actual requirement/TF

O_2 mass flow rate = standard O_2 requirement/OTE

– determine the volume flow rate of air to be blown:

O_2 volume flow rate = O_2 mass flow rate/density

air volume flow rate = O_2 volume flow rate/O_2 fraction in air (v/v)

volume air flow = O_2 mass flow/O_2 concentration in air

oxygen density = 1.428 kg/Nm^3

O_2 in air = 21% (v/v)

O_2 concentration in air = 1.428 * 0.21 = approx. 0.3 kg O_2.Nm^{-3} of air

with:

- volume flow of air in Nm^3.h^{-1};

- O_2 mass requirements in kgO_2.h^{-1};

- OTE or dimensionless oxygenation efficiency (value less than 1);

- TF: dimensionless;

- density in kg.Nm^{-3};

- O_2 fraction: dimensionless;

- the value 0.285 kgO_2/m^3 of air at 15°C should not be mistaken for 0.3 kgO_2/Nm^3 of air (at 0°C and 760 mm Hg).

6.7.7.2. *Definitions of actual requirements*

These include:

– the total daily requirement (for average pollution);

– the peak hour requirement;

– the respiration requirement.

In addition, the minimum requirement (especially in the case of aeration operating continuously) and the maximum requirement have to be defined, according to the methodology presented in volumes 2 and 3.

6.7.7.2.1. Minimum oxygen requirements

It is particularly important to take into account the minimum air flow rate when designing aerators that must operate continuously. In the case of a two-tank process, over-aeration in the second tank results in the recirculation of a super-aerated mixed liquor, preventing denitrification in the first tank.

In the case of permanent variable flow insufflation, the minimum air flow rate must take into account:

– the pollution load of the plant at start-up;

– the decrease in the pollution flow at night.

The minimum air requirement can be estimated by the following formula:

$$\text{Minimum O2 requirements (kg.h-1)} = a\ [\text{respiration requirements} + (\frac{\text{average pollution requirements}}{b})]$$

with:

– a being the loading coefficient of the plant at activation (relative to the nominal design conditions);

– b being the decrease coefficient in the flow of pollution at night. It is determined by the ratio: hourly average pollution flow/hourly minimal pollution flow;

– respiration requirements (kg $O_2.h^{-1}$), which are the biomass respiration requirements. This is a constant value over 24 hours, independent of the flow of pollution to be treated;

– average pollution requirements (kg $O_2.h^{-1}$), which are the requirements for the elimination of the average daily pollution flow.

When not outlined in the specifications, the following can be used:

– for nitrification/denitrification in raw water: b = 3;

– for nitrification/denitrification in settled water: b = 2.

6.7.7.2.2. Maximum oxygen requirements

For permanent aeration (continuous or cyclic aerator operation), aeration is not interrupted during the peak period.

For sequenced aeration, to enable denitrification during the peak of pollution, the aerators are considered to operate for a maximum of 50 minutes per hour. The aerators will therefore be designed to provide 6/5 of the air requirements of the hourly peak.

Average aerator operating time

When the aerators operate 5/6 of the time during peak periods, the average daily operating time is:

Average daily operating time = 24 x (5/6) x (average hourly requirements/peak hourly requirements)

EXAMPLE.– Assumptions:

– total daily requirements: 5,757 kg $O_2.d^{-1}$;

– respiration requirements: 2,005 kg $O_2.d^{-1}$.

We have:

peak real requirements:

peak real requirements = total daily O_2 requirements x O_2/24 peak coefficient = 5,757 x 1.39/24 = 333 (kg $O_2.h^{-1}$)

average real requirements:

average real requirements = total daily O_2/24 requirements = 5,757/24 = 240 (kg $O_2.h^{-1}$)

minimum real requirements (for a = 0.8 and b = 3):

min real requirements = a * [2,005/24 + ((5,757-2,005)/24/b)] = 108.5 (kg $O_2.h^{-1}$)

Aeration capabilities to be installed

1) *Surface aerator.*

Data:

– standard oxygen mass transfer = 1.8 kg O_2 standard.kWh^{-1};

– TC = 0.7.

If we have a permanent aeration:

peak power = real peak requirements/(1.8*0.7) = 333/(1.8*0.7) = 264 kWh

min power = minimum real requirements/(1.8*0.7) = 108.5/(1.8*0.7) = 86.1 kWh

With a sequenced aeration, peak power and operating time will be:

peak power = 333/(1.8*0.7)*6/5 = 317 kWh

average daily duration = 24*(5/6)*240/333 = 14.41 h.d^{-1}

For sequenced aeration, 317 kWh turbines must be installed, and the hourly operating time must be varied.

2) *Air insufflation.*

Data:

– standard oxygenation efficiency = 45%;

– TF = 0.5.

If we have permanent aeration:

peak air volume flow rate = 333/0.45/0.5/0.3= 4,933 Nm^3.h^{-1}

min air volume flow rate = 108.5/0.45/0.5/0.3= 1,607 Nm^3.h^{-1}

With sequenced aeration, peak power and operating time will be:

volume flow rate of air to be insufflated = 6/5*[333/0.45/0.5/0.3] = 5,920 Nm^3.h^{-1}

average daily duration = 24*(5/6)*240/333 = 14.41 h.d^{-1}

3) *Aeration operation and control.*

Aeration is usually controlled independently on each lane. In all cases, it is necessary to provide a schedule for the operation of the aerators in the event of a control failure.

6.7.7.2.3. Characteristics of aerators under standard conditions

Aeration system performance is usually specified with reference to standard test conditions, defined as follows (ASCE 2010): "tap water" as the aerated medium, water temperature 20°C, 1 atm atmospheric pressure and zero DO concentration. The system performance can then be expressed in terms of the following standard parameters:

– Standard oxygen transfer rate (SOTR, kg O_2/h).

– Standard oxygen transfer efficiency (SOTE, % oxygen absorbed).

– Standard specific oxygen transfer efficiency (SSOTE, % oxygen absorbed per meter depth).

– Standard aeration efficiency (SAE, SOTR/power input, kg O_2/kWh).

Overall liquid film coefficient, *K_La (aeration coefficient)*

The evolution of the dissolved oxygen concentration in an aerated basin with water at a temperature T is described by the equation:

$$\frac{dC}{dt} = KLaT\ (Cs - C)$$

with:

– t being time (h);

– C being the dissolved oxygen concentration at time t ($mg.L^{-1}$);

– Cs being the concentration of dissolved oxygen at saturation at temperature ($mg.L^{-1}$);

– K_La_T being the overall liquid film transfer coefficient for clean water and for a temperature T (s^{-1}).

The oxygen consumed by microbials in the wastewater during the aeration process is also known as the oxygen uptake rate (OUR). The oxygen transfer into wastewater can be determined by the overall mass oxygen transfer coefficient, K_La by using the mathematical equation:

$$\frac{dC}{dt} = K_L(Cs - C) - OUR$$

where C is the bulk liquid DO level at any time t in $mg.L^{-1}$, and Cs is the constant oxygen saturation concentrations in $mg.L^{-1}$.

Specific OC

This is defined as the mass of oxygen introduced per unit of time into the unit volume of pure water. Referring to standard values, it is the quantity in g of O_2 dissolved in one hour in one m^3 of pure water, under particular operating conditions (temperature 10°C, pressure 760 mm Hg and zero oxygen content). It is written as:

$$OC = K_La\ Cs\ (gO_2.\ m^{-3}.h^{-1})$$

Both K_L and a are a function of temperature, the former increasing and the latter decreasing. When defining the ratio of OC at any temperature to that at 20°C, the impact of temperature on this product is much less than on the oxygen transfer rate or the oxygen saturation value.

In the United States, the OC is calculated at 20°C. There is also a small difference between the two values. The shift from one to the other is made through the relation:

$$OC_{20} = OC_{10} \times 0.98$$

SOTR

The SOTR is the mass of oxygen transferred per unit time into a given volume of water and reported at standard conditions. The European literature also refers to this term as the OC. The SOTR is the decisive factor in the dimensioning of activated sludge plants. It depends on the required OUR by microorganisms.

This is the quantity of dissolved O_2 in a tank of volume V under standard conditions (clear water, 20°C, 1,013 mbar, zero dissolved oxygen concentration at time 0). It therefore expresses the mass of oxygen that can be introduced into a tank of volume V, under standard conditions:

$$\text{SOTR [kg } O_2\text{/h]} = K_L a.Cs.V.10^{-3}$$

with:

– K_La being the overall liquid film coefficient (h^{-1});

– Cs being the concentration of dissolved oxygen at saturation at temperature T ($mg.L^{-1}$);

– V being the volume of the tank (m^3).

In the case of diffused air aeration systems, Cs can exceed the standard handbook saturation concentration due to the hydrostatic pressure on the rising air bubble stream.

SAE

The SAE is the rate of oxygen transfer per unit power input, which may be based on either delivered power (DP) or wire power (WP). The overall efficiency of the aeration equipment is the product of the individual efficiencies of mechanical equipment. Typical efficiencies (EPA 1983) of the individual components are as follows: blowers (50% for older to 80% for newer units), motors (95%), coupling (95%) and gear box (95%). The SAE is used to relate the consumed wire power to that which is delivered to the air for diffused aeration or to the liquid for mechanical aeration (Mueller et al. 2002).

This is the amount of dissolved O_2 per kWh consumed. This value is commonly used in surface aerators:

SAE (kgO2.kWh-1) = = SOTR/Pcons

where Pcons is the electrical power consumed during the same period (kWh).

Net aeration efficiency (NAE)

This is the quantity of dissolved O_2 with respect to the dissipated power, measured at the shaft of the turbine or brush:

NAE ($kgO_2.kWh^{-1}$) = SAE.η (motor).η' (motor reducer), where η (motor) and η' (motor reducer) are the energy transformation efficiencies in %.

SOTE

The SOTE is the fraction of oxygen supplied to the aeration tank, which is actually transferred or dissolved into the liquid at standard conditions. It is a major design parameter for diffused aeration systems. The oxygenation or transfer efficiency enables us, depending on the air flow rate supplied by the blowers, to know the flow rate of dissolved oxygen effectively. SOTE is the basic parameter characterizing the utilized diffusers (the value is also influenced by the stirring conditions). It expresses the efficiency of aeration under standard conditions (20°C, 1.013 hPa, clean water, α,β = 1, clean diffuser 0 g O_2/m^3). It can be calculated using an oxygen mass of 0.3 kg · Nm^{-3} of air:

This value is commonly used to characterize diffusers:

$$\text{SOTE} = \frac{SOTR}{xO_2\ \rho O_2\ \text{Qair}} = \frac{SOTR}{0.3\ \text{Qair}}$$

with:

– SOTR being the hourly transfer ($kgO_2.h^{-1}$);

– Qair A being the air flow rate ($Nm^3.h^{-1}$);

– xO_2 being the O_2 volume fraction in the insufflated air (xO_2 = 20.94%);

– ρO_2 being the oxygen density ($\rho O_2 = 1.4281$ kg.Nm^{-3});

– $xO_2\ \rho O_2 = 0.30$ kg O_2.Nm^{-3} air (0°C, 1,013 mbar).

$$\frac{wo}{w} = 0.295\frac{\text{mole } O_2}{\text{mole air}}\text{x}32\frac{\text{g } O_2}{\text{g air}}\text{x}\frac{\text{mole air}}{28.964 \text{ g air}} = 0.2315\frac{\text{g } O_2}{\text{g air}}$$

As 1 kg air = 1 m^3, we can deduce:

$$\frac{wo}{w} = 0.2315 \text{ x}1.3\frac{\text{kg } O_2}{1.3 \text{ Nm}^3\text{air}} = 0.30\frac{\text{kg } O_2}{\text{Nm}^3 \text{ air}}$$

For practical purposes, the frequent use of the specific standard oxygen transfer efficiency (SSOTE) expressed in %· m^{-1}, should be noted; this practice consists of comparing a diffuser's efficiency against its submersion depth.

		SAE (standard kg O_2/kWh)				SSOTE (%/m)
		Overall average	Extreme range	Usual range	With good device	
Surface Aerators	Slow turbines	1.50	0.80–2.20	1.35–1.75	1.65	–
	Fast turbines	1.05	0.60–1.45	1.05–1.35	1.20	–
	Brushes	1.55	0.80–2.30	1.50–1.80	1.70	–
Air insufflation	Medium bubbles	1.00	0.70–1.50	0.90–1.10	1.10	2.0–3.0
	Fine bubbles	2.50	1.80–3.40	2.20–3.20	2.50	3.0–6.0
	Channel + agitation	2.80	1.20–4.25	2.50–3.80	3.00	5.0–10.0
	With static mixers	1.50	0.60–2.10	0.80–2.00	1.40	–
Differential pressure system	Pump + air uptake	0.60	0.02–1.10	0.60–0.80	0.70	–
	Pump + pressurized air	–	0.80–2.30	–	–	–

Table 6.26. *SAE and SSOTE of the main aeration systems (standard conditions at 10°C) (Duchene 2002)*

The specific or average oxygenation efficiency (% of O_2 transferred in clear water per meter of immersion of the diffusers) is used to compare the efficiency of the different insufflators.

Given the influence of the density of the diffusers, the immersion depth and the air flow rate per diffuser on the oxygenation efficiency, this parameter should be used accurately.

6.7.7.2.4. Correction factors for real conditions

The main parameters linking oxygen transfer under real and standard conditions are as follows.

Overall correction factor (TF)

The overall correction factor (TF) or K_La coefficient is defined by the following relation:

$$TF = \alpha \frac{\beta Cs\ water - C\ sludge}{Cs} 1.024^{T-20}$$

Factor α

The factor α corrects the transfer coefficient according to the nature of the interstitial liquid (pH, surfactants), the type of aerator and the SS concentration of the sludge:

$$\alpha = \frac{KLa\ sludge}{KLa\ water}$$

Aerator type	Low loading or extended aeration	Medium loading	High loading
Slow surface aerator	0.9	0.85	0.8
Fast turbine	0.9	0.85	0.8
Medium bubble insufflator	0.9	0.85	0.8
Fine bubble insufflator	0.65–0.7	0.65	0.6–0.65
Differential pressure system	0.65–0.7	0.65	0.6–0.65

Table 6.27. *Urban water factor α (average values)*

Oxygen saturation factor β

The coefficient β corrects the oxygen saturation concentration between wastewater and clean water at equal pressure and temperature, induced by the difference in ionicity between the two liquids (chlorides, sulfates and other impurities):

$$B = \frac{\text{Cs sludge}}{\text{Cs water}}$$

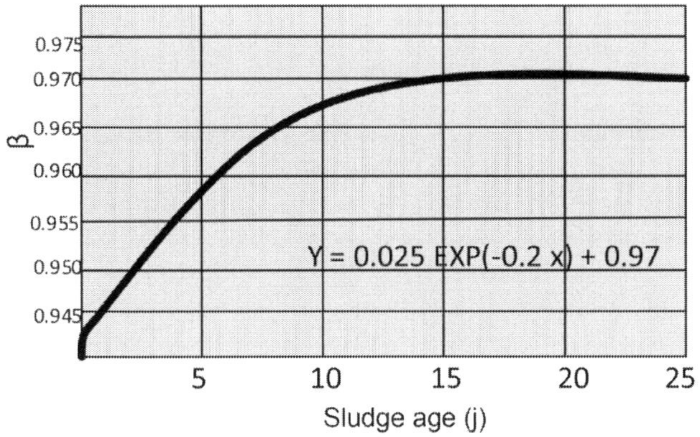

Figure 6.17. *Variation of the coefficient β according to the sludge age*

For an urban or agri-food influent, we generally have:

$$\beta = 0.95\text{–}0.97$$

Temperature influence (coefficient θ)

The variation of the transfer coefficient with respect to temperature is represented by the coefficient θ. Its value depends on the reference temperature. It is 1.024 for 20°C and 1.02 for 10°C. To express the value measured at T °C at 20°C, the following correlation can be applied:

$$K_L a_{20} = K_L a_T \times 1.024^{(20-T)}$$

Pressure influence

The concentration of dissolved oxygen at saturation varies according to the altitude H (m) with the ratio:

$$(Cs)_{TP} = (Cs)_T \, 10^{-H/18,400}$$

As an example, the saturation concentration at 10°C varies between 11.26 and 9.30 mg.L^{-1} between 0 and 1,500 m above sea level.

6.7.7.2.5. Transition from nominal to real conditions

In practice, the conditions of use of aeration devices are indeed often different from standard conditions. The working temperature is different, the nature of the liquid is variable, and even the pressure proves to be changeable.

This leads to the establishment of a corrective system according to external or relative factors of the liquid medium to be treated.

An OC has to be defined in such a way that it can be introduced into a tank of volume V at temperature T°C and pressure Pa, with dissolved oxygen content Cx (mg.L^{-1}) and finally in accordance with the conditions of salinity, suspended solids and specific surfactants. In the previous chapters, we saw to what extent K_La and Cs are corrected according to external parameters.

Oxygenation efficiency (E)

E is defined as the fraction of oxygen transferred into water due to passing one cubic meter of air.

$$\text{E (\%)} = \frac{\text{OC x Depth liquid in basin}\left(\text{m}\right)}{\text{Aeration intensity}\left(\text{ m}^3\text{ / h}\right)} x100(\%)$$

Effective oxygenation capacity

The effective oxygenation capacity (EOC) = OC x correction factors:

$$\frac{\text{EOC}}{\text{OC}} = \alpha\ 1.02^{\text{T}-10}\ \frac{\text{C}'\text{sTP} - \text{C}}{\text{Cs10}}$$

OC is the oxygenation capacity.

This is the minimal concentration that we want to maintain in the tank. It can be between 1.5 and 2 mg.L^{-1}.

C′sTP is the oxygenation capacity corrected for the working temperature and pressure. It becomes:

$$\text{C}'\text{sTP} = \frac{\text{CsT P}\ \beta}{\text{Pa}}$$

with:

$$\text{CsT} = \frac{475}{33.5 + \text{T}}$$

β relates to the suspended solids content in the water and the presence of surfactants. It is often equal to 0.95.

EXAMPLE.– An activated sludge tank must have an effective oxygenation capacity (EOC) of 60 g $O_2.m^{-3}.h^{-1}$. Assuming that the temperature is 13°C and that the minimum concentration to be maintained (Cx) is 1.5 $mg.L^{-1}$, what should be the nominal oxygenation capacity (OC) to be expected? For altitude, we take: 1,000 m; $\alpha = 0.90$; $\beta = 0.95$:

T = 13°C

Cx = 1.5 $mg.L^{-1}$

$$C'sTP = CsT\frac{9.17}{10.33}0.95$$

$$CsT = \frac{475}{33.5 + 13} = 10.21\ mg.L^{-1}$$

C′sTP = 10.21 x 0.887 x 0.95 = 8.61 $mg.L^{-1}$

$$\frac{EOC}{OC} = 0.9\ 1.02^{13-10}\ \frac{8.61 - 1.5}{10.92}$$

OEC/OC = 0.653

OC = 60/0.653 = 91.9 $g.m^{-3}.h^{-1}$

EXAMPLE.–

EOC = 60 g $O_2.m^{-3}.h^{-1}$

T °C = 18°C

Altitude = 500 m

Cx = 1.5 $mg.L^{-1}$

$$C'sTP = CsT\frac{9.17}{10.33}0.95$$

$$CsT = \frac{475}{33.5 + 18} = 9.22\ mg.L^{-1}$$

C'_{STP} = 9.22 x 0.42 x 0.95 = 8.25 $mg.L^{-1}$

$$\frac{\text{EOC}}{\text{OC}} = 0.9\ 1.02^{18-10}\ \frac{8.25 - 1.5}{10.92} = 0.65$$

OC = 60/0.65 = 92.3 g $O_2.m^{-3}.h^{-1}$

Oxygenation capacities in the presence of activated sludge

In the presence of activated sludge, the general equation for oxygen transfer involves the respiration of microorganisms. If R denotes the rate at which oxygen is consumed by the bacteria:

$$\frac{dC}{dt} = K_L a\left(Cs - C\right) - R$$

In steady state, the oxygen concentration is fixed at a value C'S which can be considered as the apparent oxygen saturation concentration in wastewater:

$$\frac{dC}{dt} = 0 = K_L a\left(Cs - C's\right) - R$$

$$K_L a = \frac{R}{Cs - C's}$$

or: R = K_La (Cs – C's)

The value of K_La includes the influence of α, temperature and the content of suspended solids. It is specific to a type of water.

Measurement of the capacity for oxygenation of aerators

The measurement protocol consists of canceling the dissolved oxygen concentration in an aeration tank filled with clean water by adding sodium sulfite and cobalt chloride. Then, we follow the evolution of the dissolved oxygen concentration when aerators are activated. This evolution of dissolved oxygen follows the equation:

$$\frac{dC}{dt} = KLa(Cs - C) \text{ or even } Log(Cs - C) = a - KLat$$

The oxygenation capacity of the clean water system is, at a temperature T and an initial dissolved O_2 concentration:

OC = K_La Cs V

with OC expressed in mg $O_2.h^{-1}$, K_La expressed in h^{-1}, Cs expressed in $mgO_2.L^{-1}$ and V expressed in L or m^3.

When OC is expressed in g $O_2.h^{-1}$, K_La is expressed in h^{-1}, Cs is expressed in $gO_2.m^{-3}$ and V is expressed in m^3.

The Log (Cs–C) curve with respect to time gives a straight line whose slope K_La leads to the OC value.

Experimental mode in clear water

The basin is previously filled with clear water. Dissolved oxygen is measured using oxygen probes, placed at different locations in the tank. These probes are carefully calibrated beforehand. The aeration system is then implemented. A cobalt chloride solution is poured into the aerator to obtain a concentration of about 1 $mg.L^{-1}$ of cobalt ion in the tank. In the same way, sodium sulphite (Na_2SO_3) is dispersed at the tank surface at a rate of 8.8 mg/mg O_2.1. This dosage is theoretically enough to maintain the dissolved oxygen concentration in the tank close to zero for at least 10 minutes. Temperature T1 and power consumption P1 are recorded during this test. The dissolved oxygen concentrations are measured at regular intervals at different points in the tank until a constant O_2 content is obtained. The temperature T2 and the power consumption P2 are registered at the end of the test.

A curve such as that shown in Figure 6.18 displaying maximum dissolved O_2 content is obtained. Considering only the dissolved O_2 values between 10% and 70% of Cs, we can then plot Log (Cs–C) with respect to time. With the slope of the line, we can extract K.

The retained temperature is the arithmetic average of T1 and T2 such that Tm = (T1 +T2)/2. The same operation is performed with power Pm.

Since the test is carried out at different measuring points, a value of K_La is determined for each sampling point. An arithmetic average of the values of K_La is established.

Knowing the Cs value at temperature Tm, it is easy to deduce the oxygenation capacity OC ($gO_2.m^{-3}.h^{-1}$) at 10°C and the atmospheric pressure.

The gross specific uptake (SAE) of clean water is obtained by:

$$\mathrm{SAE}\left(\mathrm{kgO_2 / kWh}\right) = \frac{K_L a \ \mathrm{CsTm} \ V}{P}$$

Experimental mode in the presence of activated sludge

We insulate the aeration tank by bypassing it. Similarly, aeration is stopped as well as sludge recirculation. Very quickly, the oxygen level will drop to zero due to sludge respiration. The evolution of the dissolved oxygen concentration is then monitored over time.

The equation for re-aeration is written as:

$$\frac{dC}{dt} = K_L a(Cs - C) - R$$

where R is the oxygen consumption of the sludge (mg $O_2.1^{-1}.h^{-1}$).

On the curve C according to t, a point (c, t) is chosen whose slope is p = dC/dt. The oxygenation capacity in the presence of sludge is written as:

$$OC = (p + R)\frac{Cs}{Cs - C}$$

The aeration tank is bypassed; sludge aeration and recirculation are stopped. Once the dissolved oxygen concentration is close to zero, and the various oxygen probes are properly placed around the tank, aeration is restarted. Temperature T1 and the power consumption P1 are recorded. A sludge sample is taken and transferred to a vial. It is aerated for a few minutes until its dissolved O_2 concentration is approximately 6 to 7 mg.L^{-1}. It is then placed on a magnetic stirrer. Over time, the decrease in oxygen consumed by the sludge can be read. An oxygen probe placed in the vial is used to plot C (O_2) with respect to t.

At the same time, in the tank, the dissolved oxygen values given by the probes are read at regular intervals. Aeration is extended to a constant value given by the oximeters. The temperature T'2 and the power P'2 are recorded at the end of the test. As before, an arithmetic average is made with T' and P'.

As with the clear water tests, the O_2 consumption curve in the tank is plotted according to time (Figure 6.18). The slope of the tangent at a point (C, t) gives the re-aeration rate P (mg $O_2.L^{-1}.h^{-}1$). From the results of the vial test, we can also plot the O_2 decrease curve with respect to time. The slope of this curve gives the oxygen consumption by the sludge (in mg $O_2.L^{-1}.h^{-1}$) or R.

The quantity of oxygen actually introduced per unit of time, for a concentration C, into the tank is:

$$K'_L a = V\ (R + p) \text{ in } gO_2.h^{-1}$$

The value of K'_La is transposed to standard temperature conditions.

It is assumed that the saturation concentration is identical to that in clear water, that is, Cs (10°C): 10.92 $mg.L^{-1}$.

K'_La is calculated for 10°C by the relation:

$$K'_La_{\ 10\,°C} = K'_La_{\ Tm}/1.02^{(T'm-10)}$$

The value of the coefficient α obtained by comparing the clear water and wastewater tests (vial test):

$$\alpha = K'_La \text{ wastewater} / K_La \text{ clear water}$$

This enables the correction of the effect of the medium on K_La, using the relation: K_La (10°C, clear water) = K'_La (wastewater)/α.

Finally, the oxygenation capacity is:

$$OC = (p + R)\frac{Cs}{Cs - C}$$

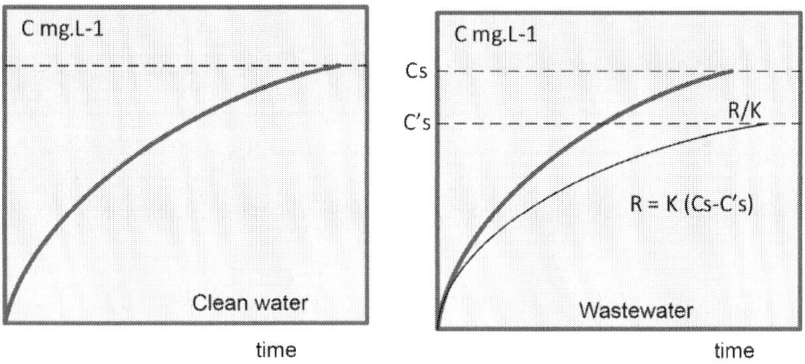

Figure 6.18. *Aeration tests in clean water and wastewater*

EXAMPLE.– *Measurement of the oxygenation capacity in clean water.* The evolution of the dissolved oxygen concentration is monitored in a tank using two probes (A and B). The results obtained are presented in Table 6.28.

The initial temperature is 18°C and the final temperature is 19°C. The consumed power is 5 kW on average. The volume of the tank is 100 m^3.

The re-aeration equation is:

$$\frac{dC}{dt} = KLa(CsT - C) \text{ or } Log\frac{CsT - Co}{CsT - C} = KLat$$

Probe A		Probe B	
C ($mgO_2.L^{-1}$)	**Time (min)**	**C ($mgO_2.L^{-1}$)**	**Time (min)**
0.7	1	0.8	1
1.1	2	1.1	2
1.7	4	1.5	4
1.9	5	1.8	5
2.4	7	2.1	7
2.9	9	2.8	9
3.2	10	3.0	10
3.6	12	3.5	12
4.0	15	4.1	15
4.9	20	4.9	20
6.0	30	6.3	30
6.6	40	6.9	40
6.9	50	7.2	50
7.1	60	7.5	60
7.4	75	7.7	75
7.4	90	7.8	90

Table 6.28. *Activated sludge re-aeration experimental results*

Co is the concentration of dissolved O_2 at the initial time (0.5 $mg.L^{-1}$). The slope of the line obtained gives K_La:

$K_La = 2.16\ h^{-1}$ (probe A)

$K_La = 2.04\ h^{-1}$ (probe B)

Cs taken at 18.5°C is equal to 9.10 $mg.L^{-1}$.

By referring to the standard values, namely 10°C:

K_La average = (2.16 + 2.04)/2 = 2.1 h^{-1}

At 10°C, we have:

$K_La_{10\,°C} = 2.1 \times 1.02^{(18\text{-}10)} = 2.48\ h^{-1}$ and $Cs_{10\,°C} = 10.92\ mg.L^{-1}$

$OC = 2.48 \times 10.92 = 27.08\ g.m^{-3}.h^{-1}$

The oxygenation efficiency is:

$27.08. \times 100 \times 10^{-3}/5 = 0.54\ kg\ O_2.kWh^{-1}$

EXAMPLE.– *Measurement of oxygenation capacity in wastewater.*

Data:

– aeration tank: volume of 100 m^3;

– aerator: one turbine with power consumption of 5 kW;

– water temperature: 15°C;

– Cs: 9.75 $mg.L^{-1}$; Co: 0.5 $mg.L^{-1}$.

C ($mgO_2.L^{-1}$)	6.4	6.3	6.1	6.0	5.8	5.8	5.5	5.3	5.0	4.3	3.8	3.5
Time (min)	0	1	2	3	4	5	8	9	10	15	18	20

Table 6.29. *Measurement of O_2 consumption in activated sludge*

From the curve C = f (t), we are able to extract R, that is, R = 0.14 mg $O_2.1^{-1}.mn^{-1}$ or 8.4 g $O_2.m^{-3}.h^{-1}$.

C ($mgO_2.L^{-1}$)	0.5	3.9	4.5	4.9	5.2	5.6	6.1	6.4	6.4	6.5	6.5	6.4
Time (min)	0	5	7	10	20	30	40	45	50	60	65	70

Table 6.30. *Reoxygenation measurement in activated sludge*

On the curve C (O_2) = f (t), we choose a point p relative to a curvature: P = dC/dt leads to p = 2.50 $gO_2.m^{-3}.h^{-1}$.

The oxygenation capacity is:

$$OC = (2.5 + 8.4)\frac{9.76}{9.76-6.1} = 29.1\ gO_2.m^{-3}.h^{-1}$$

The oxygenation efficiency is then:

$$2.91. \times 100 \times 10^{-3}/5 = 0.58 \text{ kgO}_2/\text{kWh}$$

These values are only really useful after correction to 10°C.

EXAMPLE.– *α measurement.*

Data:

– fine bubble aeration: use of porous material;

– average temperature: 20 °C;

– oxygen dosage using an oximeter;

– wastewater is synthetic, the injected organic product is peptone.

	Clean water		Wastewater	
Time (min)	**C ($mg.L^{-1}$)**	$\text{Log}\frac{Cs-Co}{Cs-C}$	**CL ($mg.L^{-1}$)**	$\text{Log}\frac{Cs-Co}{Cs-C}$
0	0.5	0	0.2	0
1	0.8	0.036	0.3	0.011
2	1.2	0.087	0.4	0.023
4	1.7	0.155	0.8	0.072
7	2.6	0.290	2.0	0.198
10	3.3	0.409	2.3	0.278
15	4.5	0.653	2.6	0.325
18	5.0	0.775	2.8	0.357
20	5.3	0.856	3.0	0.391
25	6.0	1.077	3.3	0.444
30	6.5	1.270	3.8	0.538
40	7.2	1.626	4.3	0.643
50	7.8	2.081	4.8	0.760
60	8.1	2.422	5.3	0.892

Table 6.31. *Measurement of coefficient α*

At 20°C, the oxygen saturation concentration is 8.84 $mg.L^{-1}$. In Table 6.31, C represents the value of the oxygen concentration in the tank at the same time.

At the initial time, Co = C. By plotting $\text{Log}\frac{Cs-Co}{Cs-C}$ with respect to time, it is possible to extract K_La using the slope of the line, for both clean water and wastewater.

In clean water, we obtain: $K_La = 2.55\ h^{-1}$ and for wastewater: $K_La = 1.17\ h^{-1}$.

$\alpha = 1.17/2.55 = 0.458$

6.7.8. *Using pure oxygen*

In order to avoid over-equipping conventional aeration systems, it is possible to use pure liquid oxygen, whether or not in combination with a conventional aeration system. Pure oxygen can be delivered and stored on-site or produced on-site.

This use is mainly intended for factories that have to deal with heavy seasonal pollution for a short period of time or in some cases of rehabilitation or expansion of existing units. A noticeable reduction in odors and volatile organic compounds (VOCs) can also be observed.

The most effective implementation consists of introducing oxygen through a differential pressure system coupled with water ejectors. It allows transferring up to 90% of the oxygen introduced.

Another way consists of doping the insufflated air in with liquid oxygen. Problems related to mixing the two fluids and insufficient overall efficiency would, according to suppliers, make this solution economically unsatisfactory. The lack of references, to our knowledge, prevents us from evaluating the real profitability of such an operation.

The saturation concentration of water with oxygen is $10.20\ mg.L^{-1}$ at 3°C and $10.92\ mg.L^{-1}$ at 10°C. Water can therefore only draw air at values below or equal to the saturation concentration. Assuming that the BOD of settled wastewater is between 200 and 400 mg $O_2.L^{-1}$, the maximal content to be introduced would be only the 20th or 40th part of the air required for the effective purification of the water. Aeration alone would never achieve this result unless the aeration operation were repeated 40 times, starting from a total deoxygenation point until saturation, assuming that, in the intervals, the existing aerobic bacteria proceed with the treatment operation.

Pure oxygen is sometimes used for feeding aeration systems. The most convenient way is to manufacture it on site by:

– conventional cryogenic processes;

– adsorption processes using molecular sieves, which provide gaseous oxygen while the other constituents of the air (nitrogen, CO_2, water) are blocked with molecular filters that saturate and must be periodically regenerated.

The oxygen supply is dosed according to the exact needs of the microorganisms for treatment. With this system, it is possible to maintain high oxygen levels in the tank (of about 7–8 $mg.L^{-1}$) and very high sludge concentrations. This results in reducing reaction time or increasing the applied loading. The use of pure oxygen can be considered when it is available under advantageous economic conditions, which has only been proven in industrial sites.

Pure oxygen enables:

– lower energy requirements;

– significantly increasing the mass load without the onset of bulking;

– better settling, which leads to a reduction in the size of clarifiers;

– lower sludge production;

– stability for the quality of the treated effluent;

– better toleration of sudden load variations by the reactor;

– no formation of odors and aerosols.

6.8. Activated sludge processing technology

A wide variety of activated sludge systems have been developed in recent decades to meet various treatment objectives. The principles that these processes rely on can be analyzed to explain the treatment mechanisms involved. The most common activated sludge process configurations are the following: conventional activated sludge in different configurations, contact-stabilization activated sludge, sequential biological reactors (SBRs) and membrane biological reactors (MBRs). These are explained in volume 2.

6.8.1. *Conventional activated sludge*

Conventional activated sludge systems are designed to be operated in continuous mode with biological reactors, secondary settlers and sludge extraction and recirculation equipment from the settlers to the aeration tanks. Bioreactors can be full-mixing or plug-flow bioreactors. A typical process flow diagram is shown in Figure 6.19. Conventional activated sludge is used to treat organic loading rates associated with a Q_{BOD}/X ratio of approximately 0.2 to 0.5 g $BOD_5.g^{-1}$ $VSS.d^{-1}$. The water level in reactors should be above 3 m but less than 8 m. The recommended level typically varies between 3.5 and 5.5 m. Aeration tanks should have at least an additional 0.5 m and at least 0.9 m if using aerosol-forming surface aerators.

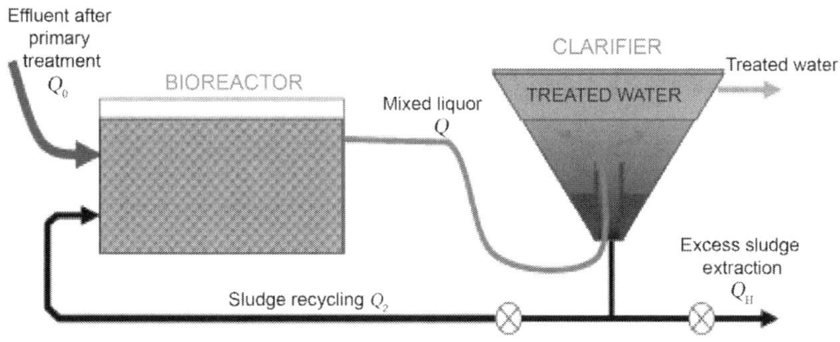

Figure 6.19. *Typical diagram of the activated sludge process*

6.8.1.1. *Fully mixed reactor*

In a fully mixed reactor, the characteristics of the liquid (MSLM concentration, BOD_5 concentration dissolved oxygen requirements, etc.) allegedly are the same throughout the reactor volume, as well as in its effluent. Proper mixing in the tank depends on several factors, including the length-to-width ratio, the mixing capacity and the location of the influent feeding and effluent discharge structures. In practice, to be considered fully mixed, the reactor would have to have a length-to-width ratio of less than or equal to 3:1 (WEF 2018). The dimensions of the tank should also be adjusted according to the mixing equipment (surface aerators, turbines or air diffusers). This type of reactor offers the following advantages: good resistance to peak loads and toxic shocks without significant repercussions on the quality of the effluent and simplified aeration design compared to a plug-flow reactor since the oxygen demand is theoretically the same everywhere.

Fully mixed reactors have a uniform Q_{BOD}/X ratio throughout the basin, which makes them more prone to filament issues than plug-flow reactors, where a decrease in this ratio is observed along the reactor.

6.8.1.2. *Oxidation channel*

The channel is a tank with an oblong or circular shape. Suitable equipment ensures the circulation of the water to be treated between the inner and outer walls. In a channel, the flow is plug flow, and the concentrate distribution is of the fully mixed type. The oxidation channel is a variant of the activated sludge process at low load. In this respect, the retention times and the sludge age achieved enable a thorough removal of carbon and nitrogen compounds. The channel is thus defined as a "single-sludge wastewater treatment system". Channels can be classified according to their shape. In oblong channels, aerators can be placed horizontally (brushes) or vertically (turbines). The latter case is known as a carousel channel. Circular channels include a central area that can act as an anaerobic basin. In small oxidation channels, alternating the aeration is a fairly suitable solution for nitrogen and phosphorus removal. The pre-treatment stage is still present upstream of this unit. In contrast, primary settling is not always necessary. This treatment system includes sludge clarification and recycling. Optionally, an anaerobic tank can be added to the treatment system to ensure the biological removal of phosphate from water. Stirring and aeration are carried out using wire brushes called "mammoth" brushes. The mass of water circulates in a longitudinal fashion and is animated by a continuous rotational movement. The brush ensures sufficient oxygenation capacity that allows satisfactory BOD removal. With stirring, microorganisms are brought into contact with incoming pollution, on the one hand, and the oxygen necessary for oxidation is provided, on the other hand. A plug-flow reactor is a long, narrow tank that can be likened to multiple fully mixed reactors in series. To obtain an ideal plug-flow reactor, the length-to-width ratio should be 10:1 (Metcalf & Eddy, Inc. and Aecom 2014; WEF 2018). However, to avoid very high downtime costs, a length-to-width ratio of 4:1 (MOE 2008) or 5:1 is acceptable. The width of the reactor is usually 5 to 9 m (WEF 2018). The advantages of oxidation channels reside in their ability to absorb hydraulic and biological shocks. The fairly long retention times allow them to control hydraulic shock and sudden increases in loads. The sludge production is lower compared to other treatment systems due to the long retention time, and there is high oxidation of the organic matter.

For a plug-flow tank, the feeding flows from the inlet to the outlet with a piston effect. As a result, BOD undergoes decreasing degradation from the inlet to the outlet, and the mass load is greater at the inlet than at the outlet.

Figure 6.21 (in the next section) shows a diagram of a full treatment plant.

In a plug-flow reactor, the feeding is carried out in the inlet area with water highly loaded with organic matter. The S/M ratio decreases as the water travels through the reactor and the organic matter is consumed. This aspect hampers the growth of most types of filamentous bacteria, which ensures better settling of the produced sludge. This type of reactor offers the following advantages:

– good resistance to peak flows without significant impact on effluent quality (WEF 2018);

– better sludge settling and more efficient nitrification than a fully mixed reactor.

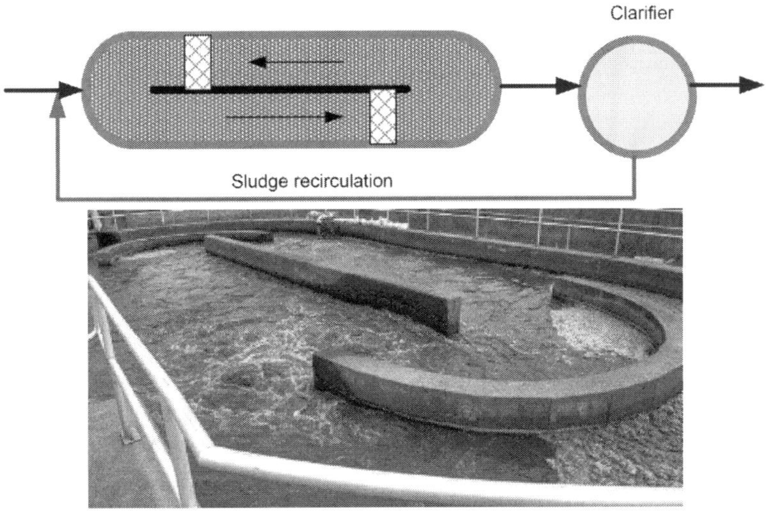

Figure 6.20. *Oxidation channel operating principle*

6.8.1.2.1. Oxidation channel with compartmentalizations

Wastewater and return sludge are admitted at the same time and travel in a block in the tank. Pollutant removal, O_2 demand and sludge condition vary throughout the unit along the path of the wastewater.

In a plug-flow tank, the pollution concentration decreases as the water flows into the tank. This plug flow is generally provided by baffles in a rectangular tank, and each compartment operates as a full mixing tank. In a tank without baffles, one has a plug flow when the length is more than four times the width. The concentration gradient thus obtained makes it possible to limit the bulking risks.

The alternating of aerobic and anoxic zones allows for the simultaneous treatment of carbon and nitrogen pollution. Since the recirculation time in the channel is small compared to the retention time, the medium of the biomass rapidly changes, which stimulates the growth of several types of microorganisms. Heterotrophic and autotrophic bacteria are active in the aerated zone. Denitrifying bacteria grow in the anoxia zone. The channel is capable of a fairly high degree of mixing. This mixing is made possible by the circulation time and the turbulence caused by the brewing. Thereby, it allows coping with hydraulic shock.

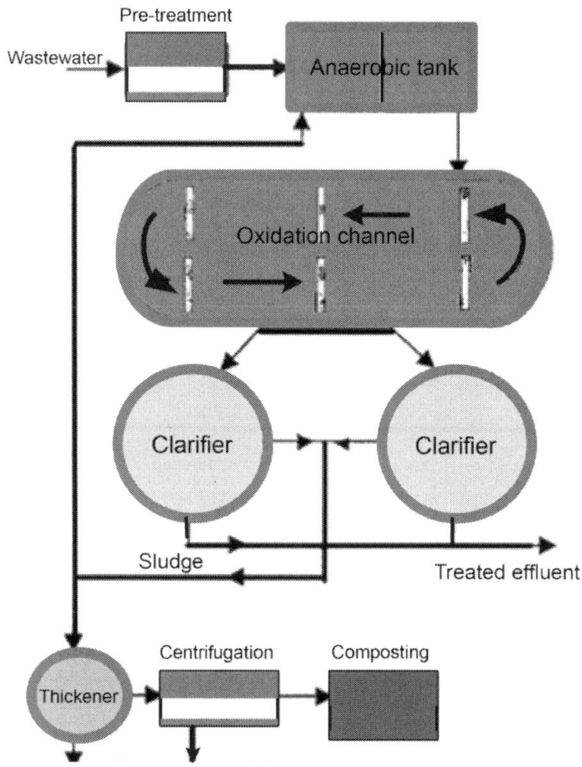

Figure 6.21. *Single-feed oxidation channel*

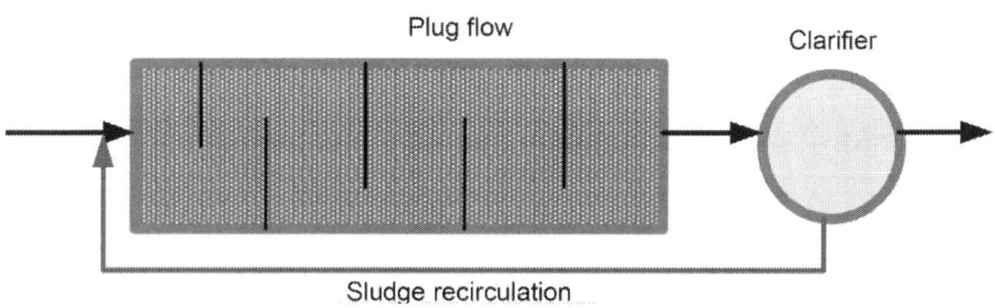

Figure 6.22. *Oxidation channel with compartmentalization*

It is recommended that this type of channel be implemented for the treatment of carbon and nitrogen pollution.

6.8.1.3. *Gradual feeding*

The gradual feed channel distributes the wastewater at several points along its length, as shown in Figure 6.23. This configuration tends to make the oxygen demand uniform throughout the plug-flow reactor (WEF 2018). The recirculated sludge is always introduced into the reactor head. The various points that supply wastewater dilute the mixed liquor, which reduces the solids loading rate at the clarifier inlet. A higher hydraulic loading rate can therefore be applied at the secondary settler to reduce the risk of organic overload and mitigate poor performance. This configuration makes it possible to adjust the applied Q_{BOD}/X ratio by shifting the feed upstream or downstream.

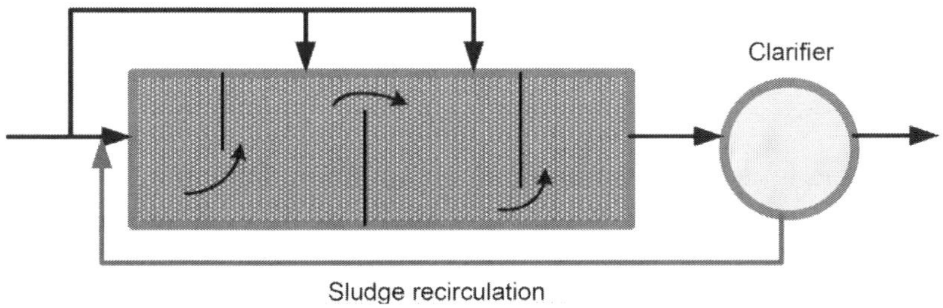

Figure 6.23. *Gradual feeding*

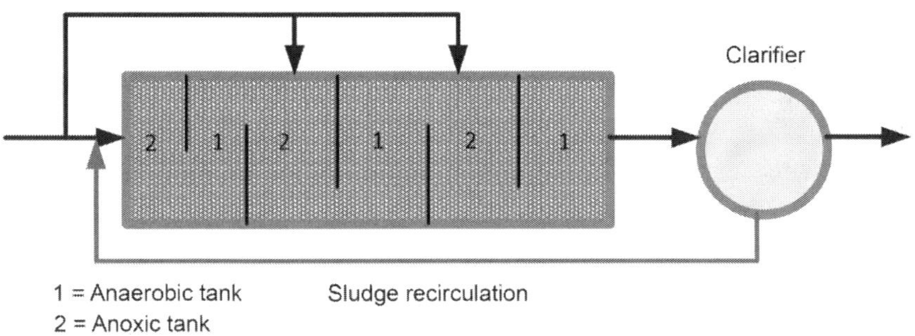

Figure 6.24. *Gradual feeding plug flow including aerobic and anoxic compartments*

For nitrogen removal (see volume 2), with gradual feeding, the anoxia compartments are supplied with raw water, that is, with a high concentration of organic matter that enables the acceleration of the denitrification kinetics. The sludge travels following a plug flow, but the effluent is admitted at different points in the tank. This reduces the hydraulic retention time or increases the loading rate of the tank.

6.8.2. *Contact stabilization*

The contact stabilization biological reactor (CSBR) process is a variant of activated sludge wastewater treatment. It has been developed to take advantage of the absorbent properties of the activated sludge process. BOD is removed from the activated sludge process in two phases. This process is as follows:

– The pollution is adsorbed, in the form of nutrient reserves, in a first aerated tank, known as a contact tank.

– After passing through the settler, the sludge is sent to an aerated stabilization tank where the reserves are consumed.

In a CSBR system, the process is divided into two main stages, the contact stage and the stabilization stage:

1) Contact stage or absorption phase: during this step, the raw wastewater is mixed with a flow-back of activated sludge in a contact tank. This is a short stage and aims at the rapid absorption of soluble organic pollutants by microorganisms present in the sludge. During this phase, most of the finely divided, colloidal and dissolved organic suspended matter is absorbed by the activated sludge. This takes 30 to 40 minutes. The objective is to maximize the adsorption and absorption of organic matter by microorganisms. The contact phase is usually followed by primary settling, which separates the sludge from the treated water. The so-called contact tank is not a bacteria-selective tank. This is a reserve adsorption tank. For the removal of carbon and nitrogen pollution, all the tanks must operate at low load or extended aeration.

2) Stabilization phase or oxidation phase: during this phase, the oxidation of organic matter occurs. The separated sludge, rich in organic matter, is then conveyed to a stabilization tank where it is kept for a longer period of time, allowing microorganisms to degrade these materials. This step makes it possible to biologically stabilize the sludge by reducing the organic load. After stabilization, the sludge is settled in a secondary clarifier. The clear part is discharged as treated effluent, while the excess sludge is removed from the system, and some is recirculated back to the contact phase. The flow-back sludge, pre-aerated for several hours, is admitted into the tank at the same time as the effluent. Pollution adsorption is then maximal and decisive: 1 hour of contact can remove 90% of the BOD_5. Reactivation then enables assimilation. It lasts only a few hours.

The two phases are separated and take place in two separate tanks as part of the contact stabilization process. During sludge reaeration, the absorbed organic matter is oxidized to restore adsorption energy and produce new cells. In this case, the required aeration volume is about 50% of that of a conventional process.

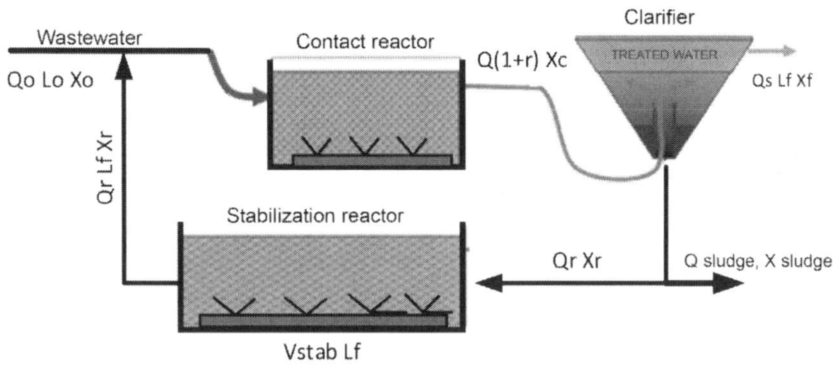

Figure 6.25. *Contact stabilization*

The contact-stabilization or contact-assimilation process results in limiting the risks of bulking when treating an influent with an agri-food component containing in particular sugary substances, where the adsorbable soluble COD/total COD ratio is high. It can be used as a treatment of carbon pollution alone or for nitrification-denitrification. The role of the contact zone is to anticipate the risks of degradation of the quality of the activated sludge defined by the MI resulting from a nutritional imbalance in the aeration tank and conducive to the development of filamentous bacteria, which consequently leads to poor settling and significant SS leakage.

We have:

- Qrcz: recirculation rate in the contact zone;
- tc: contact time, 10–15 min for Qpdt and 20–25 min for Qadt;
- Qpdt: peak flow rate dry time;
- Qadt: average flow rate dry time;
- Xr: recirculated sludge concentration ($g.L^{-1}$);
- absorbable COD ($mg.L^{-1}$): 20–50% raw influent COD;
- COD load in mg.g of recirculated sludge: 80–130 depending on the mass load;
- contact zone volume:

$$Vzc = (Qrcz + Qpdt)\ tc/60$$

Contact stabilization is quite effective for wastewater treatment in general but not very effective for industrial wastewater treatment, where the organic matter present in the wastewater is mostly in dissolved form. Existing treatment plants can be modified by changing the piping and providing a partition in the aeration tank. This upgrade can increase the capacity of the existing facilities. Contact stabilization is also effective for wastewater treatment involving the presence of organic matter in colloidal form. It may not be as effective for treating wastewater containing organic matter only in soluble form.

The CSBR process is particularly advantageous when there are considerable variations in the pollutant load or wastewater flow, as it is capable of greater resilience to load shocks and enables better sludge management. It is often used in facilities where space is limited, as it requires less reactor volume compared to the conventional activated sludge system. Products or systems that may be associated with this process include equipment such as fine bubble air diffusers, floating fast aerators or aeration turbines, which ensure efficient sludge oxygenation during the contact and stabilization phases. This equipment contributes to the efficiency of the process through its capability to provide adequate oxygenation and homogeneous mixing, thus inducing biological activity and the degradation of organic pollutants.

6.8.3. *Full mixing*

In this system, wastewater is introduced at the reactor end, and the effluent is discharged at the other end (Figure 6.26). Recirculated sludge can be mixed with the influent upstream of the reactor or sent separately to the reactor. If the recirculated tributary and sludge lines are separated, it is easier to convert the conventional bioreactor feed to another type of feed when upgrading the wastewater treatment plant. When the influent and the recirculated sludge are combined upstream, it is important to ensure good mixing, especially if the flow is directed to several bioreactors in parallel, as their respective viscosities are slightly different. In this type of tank, there is almost absolute homogeneity at the BOD concentration or sludge concentration level. The sludge and effluent are therefore finely mixed. Pollutants, sludge and O_2 demand are uniformly distributed. This is the most widespread system. The mass load is uniform, and reaction times are short. The basin can be rectangular or circular, the circular shape being more advantageous in terms of civil engineering but occupying more space.

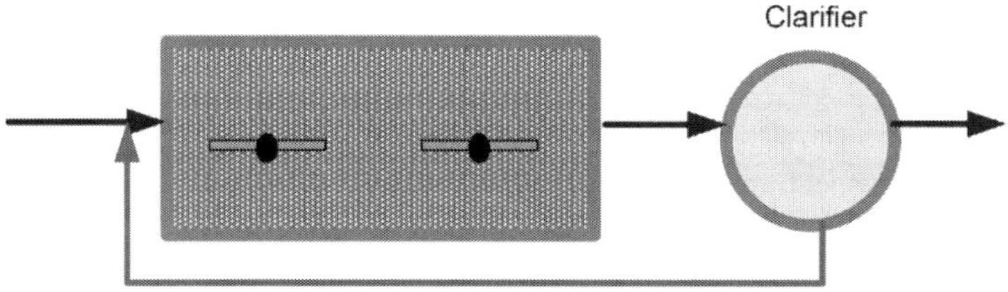

Figure 6.26. *Fully mixed biological tank*

The removal of carbon pollution alone is carried out in a medium- or high-load tank. Permanent aeration (aerobic conditions maintained) is carried out with continuous or cyclical operation of the aerators. The O_2 consumption is very fast. In cyclic operations, the admissible frequencies for reinitiating the aerators must be respected. This problem is amplified if the aerators are over-dimensioned or if the station is underloaded.

6.8.4. *Sequencing batch reactor*

Sequencing batch reactors (SBRs) are a variant of the activated sludge process. Their particularity is that biomass settling is directly achieved in the aeration tank rather than in a separate settler. The bioreactor typically operates in batch mode following a sequence that typically includes the following steps: filling, reaction, settling, supernatant removal and resting (Figure 6.27).

The sludge can be drawn off during the various stages except during filling. Since the flow of wastewater reaching the treatment plant is variable and continuous, a buffer tank should be allocated to store the influent or several reactors in parallel such that one of them is assigned for refilling (WEF 2018). Some SBRs can also be designed to be continuously fed. The volume of the tank should be determined from a mass balance based on biological kinetics, taking into account variations in water level within a cycle. It is recommended that a minimal concentration of at least 2,000 mg/L of MVLSS be targeted.

Once the total required volume has been determined, the number of reactors as well as the duration of an average cycle and of each cycle step must be defined. When several reactors are involved, it is preferable that they have identical dimensions. Unlike other activated sludge processes, the design of batch SBRs is based on the maximal daily flow rate and not on the average daily rate. The volume allocated to the influent during the settling stage should be enough to admit 75% of the maximal daily flow with the largest inactive unit, without changing cycle times (MOE 2008).

On the contrary, the duration of each cycle stage may require adjustment relative to operation at the average flow rate. For example, the total aeration time can be extended to ensure proper removal of organic matter. The settling time must still be sufficient. The volume should also provide a minimal hydraulic retention time (HRT) of 4 h at maximal daily flow (100%) despite the inactivity of the largest reactor (MOE 2008).

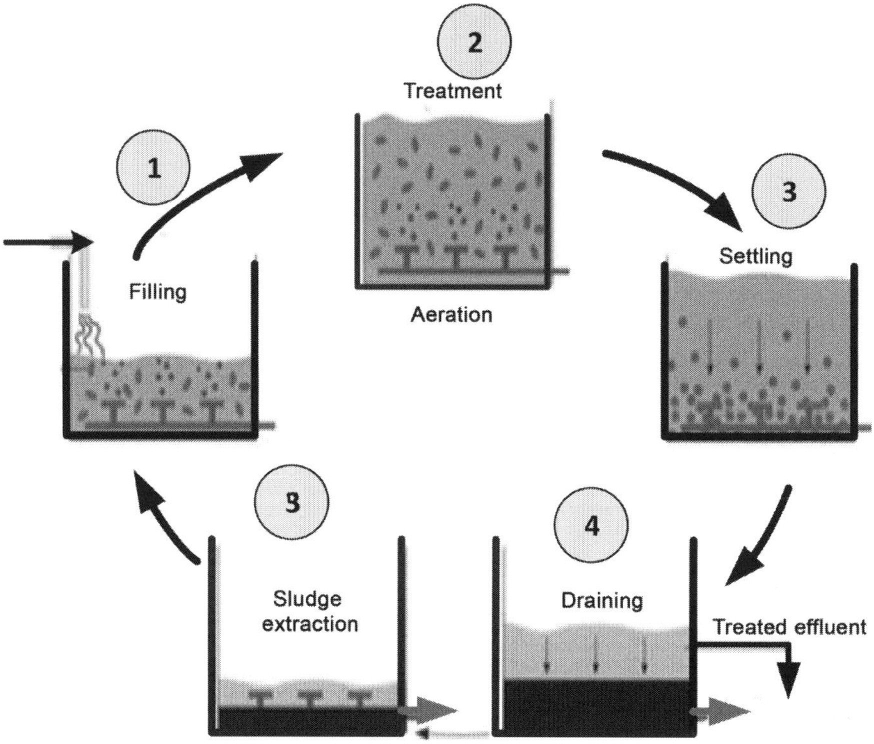

Figure 6.27. *Sequencing batch reactor (SBR)*

The average duration of the steps can reach 1.5 h to 3 h for the reaction, 0.5 h to 1 h for settling and 1 hour for extraction. Rest time is a safety margin for periods of higher flows, especially when hydraulic variations are difficult to anticipate. To verify that the reaction time allows the organic matter to be degraded, the following equation can be used:

$$KsLn\frac{Lo}{Lf} + (Lo - Lf) = Xa\,\frac{\mu max}{am}t\;reaction$$

with:

– Ks being the substrate half-saturation constant (mg substrate.L^{-1});

– Lo being the concentration of carbonaceous substrate in the influent (mg.L^{-1});

– Lf being the concentration of carbonaceous substrate in the effluent (mg.L^{-1});

– Xa being the concentration of heterotrophic active biomass in the mixed liquor (mg $VSS.L^{-1}$);

– μmax being the maximum specific growth rate of heterotrophic bacteria (d^{-1});

– am being the heterotrophic biomass synthesis coefficient (mg $VSS.mg^{-1}$ substrate);

– t reaction being the duration of the reaction step (d).

The effluent withdrawal system should have the capacity to discharge 75% of the maximal daily flow with one non-operating unit while maintaining the same cycle time (MOE 2008), and the maximal feed velocity at its inlet should not exceed 3 m/s. It should be designed to prevent the entrainment of scum or floating matter. A system should also be provided that allows redirecting the mixed liquor trapped in the discharge device during the reaction stage to prevent the degradation of the effluent. Sludge extraction can be carried out at the reaction, withdrawal or resting stages. From an operational point of view, the sludge should be preferably removed at the reaction stage, because its solid concentration is constant, uniform (quantity of fine particles and large flocs) and equal to the concentration of MVLSS. Consequently, the operator can have better control of the sludge retention time (SRT). Dryness is, however, less than when the sludge is withdrawn after the settling stage. The manufacturer should design the solid units such that the operator will be able to select the most appropriate extraction cycle. The manufacturer must ensure that the capacity of the aeration system is sufficient to provide the oxygen necessary for the degradation of organic matter and nitrification during the aerated stages of the operating cycle. In addition, as the substrate concentration decreases over time, the oxygen demand is highest at the beginning of the cycle and decreases thereafter. Therefore, the aeration system should be designed to accommodate variations in oxygen requirements (Metcalf & Eddy, Inc. and Aecom 2014). It is also recommended to include a mixing step with no aeration for denitrification, if needed. In such a case, the aeration system could be a jet aerator or fine or large bubble aerators combined with mechanical mixers. The system should be able to fully mix the reactor within five minutes from a settled state without the aid of aeration (MOE 2008).

6.8.5. *Bioring™ process*

This process has been optimized and equipped with the latest technology to provide extremely cost-effective wastewater treatment, for which it is well suited. It recreates the concept of activated sludge. The process consists of inciting the development of biomass in a tank fed by settled wastewater. Latent bacterial colonies present in raw wastewater use pollution as a nutrient for their growth and absorb it. Aeration not only mixes the mixed liquor (namely settled wastewater + biomass), it also provides the oxygen needed to keep the bacteria alive and allow for their normal growth rate. Given the quality requirements for the discharged effluents, the process must not only remove a significant amount of carbon pollution (COD or BOD) and suspended solids (SS) but also be efficient in terms of nutrient removal (nitrogen, N and phosphorus, P). Bioring™ comprises several tanks operating in series, each with specific dimensions and equipment to perform the treatment functions (Figure 6.28).

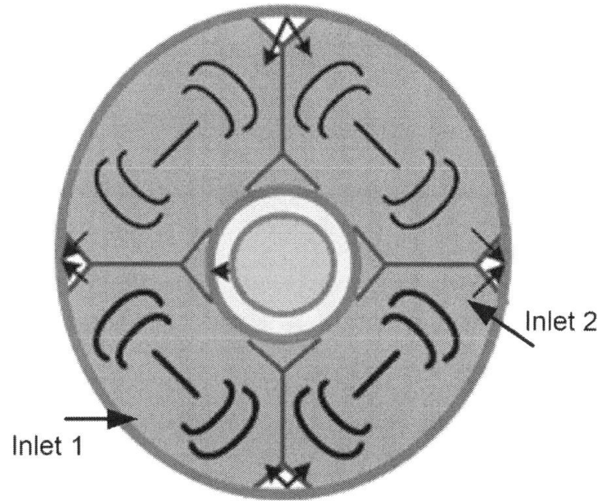

Figure 6.28. *Bioring™ schematic diagram*

In each lane, the compartments communicate through their automatic weirs, which are available to inflowing wastewater and also to the treated outflowing effluent. Each tank is equipped with aeration and mixing equipment. The nitrification and denitrification processes operate separately and alternately in each of the tanks, so there is no need to force the recirculation of the mixed liquor.

In order to achieve the required level of nitrogen removal throughout the Bioring nitrogen removing (BNR) process, under normal circumstances, the process will be operating in Biodenipho™ mode. In this case, the "activated sludge" mode works in an alternating manner. If, for some reason, a tank were to be deactivated and put out of operation, the process would operate with three independent tanks.

The BNR system in both cases will produce an effluent that will have an ammoniacal nitrogen concentration < 1 mg.L^{-1} and a total phosphorus concentration < 1 mg. L^{-1} at the outlet.

Figure 6.29. *Bioring™ at the Illawarra plant (Australia)*

6.9. Examples of dimensioning the activated sludge process

6.9.1. *Example 1: calculation of a low load aeration tank*

Effluents with the following characteristics must be treated at low loads:

– flow rate: 1,000 $m^3.d^{-1}$;

– inlet BOD_5: 360 $mg.L^{-1}$;

– SS influent: 250 $mg.L^{-1}$;

– SS effluent: 30 $mg.L^{-1}$;

– discharge effluent: BOD_5 of 40 $mg.L^{-1}$ and 30 $mg.L^{-1}$ of SS. The parameters are: a' = 0.55; b' = 0.1; am = 0.75; f = 0.3; k' = 0.4;

– sludge concentration in the tank (in MVLSS): 3 $g.L^{-1}$.

In the phase of slowed growth, the retention time is derived from:

$$\frac{Lf}{Lo} = \frac{1}{1 - k'Xat}$$

$$\frac{40}{360} = \frac{1}{1 - 0.4 \times 3 \times t}$$

t = 6.68 h # 7 h

The volume of the tank is:

$$V = Q\,t = (1{,}000 \times 7)/24 = 291.6\ m^3$$

The volumetric loading is therefore:

$$Cv = \frac{(360 - 40)1{,}000\ 10^{-3}}{290} = 1.10\ kg\ BOD_5.m^{-3}.d^{-1}$$

The mass loading is:

$$Cm = (0.360 - 0.040) \times 1{,}000 = 0.36\ kg\ BOD.kg^{-1}\ VSS.d^{-1}$$

The oxygen requirements become:

$$qO_2/t = a'\ Le + b'\ Xt$$

$$qO_2/t = (0.55 \times 320) + (0.1 \times 290 \times 3)$$

$$qO_2/t = 263\ kg\ O_2.d^{-1}$$

which leads to 263/320 = 0.82 kg $O_2.kg^{-1}$ removed BOD_5.

With a medium bubble aeration system, yielding an apparent efficiency of 7% at an insufflation depth of 4 m, we should have:

$$263/0.07 = 3{,}757\ kg\ O_2$$

which is an air flow rate of 13,400 $m^3.d^{-1}$.

The production of sludge is thus established as follows:

$$\Delta Xt = (Xmineral + f\ Xhard - Xeffluent) + amLe - bXt$$

$$X\ mineral = 1{,}000 \times 0.250 \times 0.35 = 87.5\ kg.d^{-1}$$

It is estimated that raw water contains about 65% organic matter, 30% of which is not biodegradable (f = 0.3):

$$f\ Xhard = 1{,}000 \times 0.3 \times 0.65 \times 0.25 = 21\ kg.d^{-1}$$

$Xeffluent = 1{,}000 \times 0.03 = 30\ kg.d^{-1}$

$am\ Le = 0.75 \times 320 = 240\ kg.d^{-1}$

$b\ Xt = 0.07 \times 870 = 60.9\ kg.d^{-1}$

The sludge production is therefore equal to:

$\Delta Xt = (87.5 + 48.75 + 240) - (87 + 30) = 200.2\ kg.d^{-1}$

which is 0.81 kg VSS. kg^{-1} removed BOD_5.

6.9.2. *Example 2: calculation of an aeration tank fed under medium loading*

We want to dimension the activated sludge treatment unit for a city of 6,000 inhabitants. A daily consumption of 200 L/person.d will be assumed. The values of the parameters determined in the laboratory are as follows:

– a' = 0.54; b' = 0.08; am = 0.70; k' = - 1.20;

– inlet $BOD_5 = 280\ mg.L^{-1}$;

– SS = 1,200 $mg.L^{-1}$ at 80% organic matter.

The flow rate of water to be treated is:

$Q = 200 \times 6{,}000 = 1{,}200{,}000\ L.d^{-1} = 1{,}200\ m^3.d^{-1}$

When aiming at an effluent treated at 30 $mg.L^{-1}$ in BOD_5, the treatment efficiency must be:

Efficiency = (280 - 30)/280 = 89.2%

If an MVLSS of 3 $g.L^{-1}$ is assumed in the tank, the water retention time can be determined as:

$$\frac{Lf}{Lo} = \frac{1}{1 - k'Xat}$$

$$\frac{30}{280} = \frac{1}{1 - 1.2 \times 3 \times t}$$

t = 2.32 h

The volume of the tank is thus:

$$V = Q\ (m^3/h) \times t$$

$$V = 50 \times 2.32 = 116\ m^3$$

The mass of BOD_5 to be removed is (280 - 30) x 1,200 = 300 $kg.d^{-1}$.

The volumetric load charge is then equal to:

$$Cv = 300/116 = 2.58\ kg\ BOB_5.m^3.d^{-1}$$

The mass loading is:

$$Cm = 300/(116 \times 3) = 0.86\ kg\ BOD_5.kg\ MVS^{-1}.d^{-1}$$

The calculation of Cm is often established by assuming that 70% of the sludge is in the aerator, and 30% in the settler:

$$Cm = 300/(116 \times 3/0.7)] = 0.60\ kg\ BOD_5.kg\ MVS^{-1}.d^{-1}$$

Oxygen requirements can be evaluated from the relation:

$$q\ O_2/t = a'\ Le + b'\ Xt$$

$$q\ O_2/t = 0.54 \times 300 + 0.08 \times 116 \times 3$$

$$q\ O_2/t = 162 + 27.8 = 189.8\ kg\ O_2.d^{-1}$$

or 0.63 kg $O_2.kg^{-1}$ removed BOD_5.

With a medium bubble aeration system, yielding an apparent efficiency of 7% at an insufflation depth of 4 m, we should have:

$$189.8/0.07 = 2{,}711\ kg\ O_2$$

which is an air flow of 9,678 $m^3.d^{-1}$.

The mass of excess sludge can be deduced from the relation:

$$\Delta Xt = (Xmineral + f\ Xhard - Xeffluent) + amLe - bXt$$

$$Xmineral = 1{,}200 \times 0.10 \times 0.20 = 24\ kg.d^{-1}$$

$$f\ Xhard = 200 \times 0.80 \times 0.10 \times 0.8 = 76.8\ kg.d^{-1}$$

$Xeffluent = 1{,}200 \times 0.03 = 36\ kg.d^{-1}$

$am\ Le = 0.70 \times 300 = 210\ kg.d^{-1}$

$b\ Xt = (0.08 \times 348)/1.42 = 19.6\ kg.d^{-1}$

$\Delta Xt = (24 + 76.8 + 210) - (36 + 19.6) = 255.2\ kg.d^{-1}$

or 0.85 kg MVS.kg^{-1} removed DBO_5.

6.9.3. *Example 3: population of 3,000,000 inhabitants*

Domestic wastewater

– Per capita consumption: 200 L.d^{-1}.

– Q24 = 3,000,000 x 200/1,000 = 600,000 m^3.d^{-1} = 6.94 m^3.s^{-1}.

– Q max = 6.94 x 1.33 = 9.23 m^3.s^{-1}.

– Qd = 6.94 x 1.18 = 8.18 m^3.s^{-1}.

– Q night: 6.94 x 0.87 = 6.03 m^3.s^{-1}.

– Q wet weather = 2 Q max = 18.46 m^3.s^{-1}.

Wastewater lifting station (Archimedes' screw)

Lifting station 1

– Peak flow rate: 12.55 m^3.s^{-1}.

– Elevation height: 7.50 m.

– Choice:

- two screws of 2.3 m^3.s^{-1};
- two screws of 4.3 m^3.s^{-1}.

– Engine power:

$$PM = \frac{12{,}550\ \left(\frac{l}{s}\right) 7.50\ \text{m}}{0.65 \times 102} = 1{,}420\ \text{kW}$$

Lifting station 2

– Peak flow rate: 18.46 $m^3.s^{-1}$.

– Elevation height: 5.20 m.

– Choice:

- four screws of 2.615 $m^3.s^{-1}$;
- two screws of 4.0 $m^3.s^{-1}$.

– Engine power:

$$PM = \frac{18{,}460 \left(\frac{l}{s}\right) 5.20 \text{ m}}{0.65 \text{ x } 102} = 1{,}447 \text{ kW}$$

Grit removal

– Fine screen:

$$L = \frac{d + e}{e} \frac{1}{1 - \eta} \frac{Q}{v\,h}$$

with:

- L being the total screen width (m);
- D being the bar width (15 mm);
- e being the spacing between bars (20 mm);
- η being the degree of soiling (20%);
- Q being 18.46 $m^3.s^{-1}$;
- h being the water level upstream of the bar racks (m);
- V being the flow velocity through the bar racks (1.2 $m.s^{-1}$):

$$L = \frac{0.015 + 0.020}{0.020} \frac{1}{1 - 0.20} \frac{18.46}{1.2 \text{ x } 1.8}$$

L = 18.70 m

which gives six fine screens of 3.20 m.

– Coarse screens: choice of six coarse 3.20 m screens with a 60 mm spacing between bars.

Roughing tank

– Transit time: t = 0.5 h (at Q max).

– Volume required: 9,230 x 3.6 x 0.5 = 16,614 m^3.

– Four circular 41 m tanks.

– Average depth: 3.15 m.

– Available volume: $\frac{3.14\,(41)^2}{4}$ 3.15 x 4 = 16,626.8 m^3.

– Available area: 16,626.8/3.15 = 5,278 m^2.

– Surface loading: 9,230 x 3.6/5,278 = 6.30 $m^3/m^2/h$.

– Quantity of fresh sludge:

  - 50% of settleable materials are recovered in the roughing tanks;
  - 60 g dry solids/person/d x 0.50 = 30 $g.person^{-1}.d^{-1}$ at 2.5% of dry solids;
  - 3,000,000 person x 0.030 kg dry $solids.person^{-1}.d^{-1}$ = 90,000 kg dry $solids.d^{-1}$.

– Additional rainwater treatment: 30,000 kg $MS.d^{-1}$.

That is, in total: 120,000 $kg.d^{-1}$.

Or further: 120,000 kg $SM.d^{-1}/25\ kg/m^3$ = 4,800 $m^3.d^{-1}$.

Biological basins

– BOD_5 loading: following a removal of 50% of settleable matter, 20 g $BOD_5.person^{-1}.d^{-1}$ x 0.50 = 10 g $BOD_5.person^{-1}.d^{-1}$ are retained in roughing tanks:

- biological basins therefore receive:

  60 – 10 = 50 g $BOD_5.person^{-1}.d^{-1}$

- BOD_5 after roughing: 3,000,000 persons x 0.050 = 150,000 kg $BOD_5.d^{-1}$;
- BOD_5 from sludge treatment flowback water: 10,000 kg $BOD_5.d^{-1}$;
- total BOD_5: 160,000 kg $BOD_5.d^{-1}$.

– Design:

- volumetric loading: Cv = 1.0 kg $BOD_5.m^{-3}.d^{-1}$;
- sludge dry matter in tanks: Xb = 3.3 kg dry solids (DS) m^{-3};
- mass loading: Cm = 0.30 kg $BOD_5.kg\ DS^{-1}.d^{-1}$;

- Mohlman's index: MI = 150 $mL.g^{-1}$;
- sludge volume: Vb = 500 $mL.L^{-1}$.

– Volume of biological basins:

- total BOD_5/Cv = 160,000 (kg $BOD_5.d^{-1}$)/1.0 (kg $BOD_5.m^{-3}.d^{-1}$) = 160,000 m^3;
- that is, eight basins: L = 90 m, l = 45 m, P = 5 m;
- available volume: 162,000 m^3;
- volumetric loading, Cv = 160,000 (kg $BOD_5.d^{-1}$)/162,000 (m^3) = 0.99 (kg $BOD_5.m^{-3}.d^{-1}$);
- hydraulic head: C_h = 6.94 x 86,400/162,000 = 3.68 $m^3.d^{-1}$;
- aeration time, calculated on Q_{24}: 162,000 (m^3)/(6.94 x 3.6) ($m^3.h^{-1}$) = 6.5 h;
- activated sludge recirculation ratio.

Concentration of activated sludge solids in recirculation:

$$Xr = \frac{1{,}200}{MI} = \frac{1{,}200}{150} = 8 \text{ kg DS.m}^{-3}$$

Recirculation ratio:

$$Rr = \frac{Xb}{Xr - Xb} 100 = \frac{3.3}{8 - 3.3} 100 = 70\%$$

Recirculated activated sludge flow rate: 6.94 x 0.70 = 4.86 $m^3.s^{-1}$.

The recirculation pumps are designed for a 100% recirculation ratio or 6.94 m^3/s. Two Archimedes' screw lifting stations were installed with a unit flow rate of 3.47 $m^3.s^{-1}$; that is, two screws with a unit flow rate of 1.74 $m^3.s^{-1}$.

– Elevation height: 2 m.

– Engine power:

$$PM = \frac{2(\frac{m^3}{s}) \text{ x } 3{,}600 \text{ x } 2.00(m)}{0.65 \text{ x } 102} = 218 \text{ kW}$$

– Excess sludge production:

$$\Delta Xb = Ch\,[\,0.6\,Co + 0.6\,Xo - Xs] - b\,Xb$$

a_m = 0.60

– BOD_5o: BOD_5 concentration at the biological basin inlet:

$$BOD_5o = \frac{160{,}000\ kgDBO5/d}{6{,}940\ x\ 3.6\ x\ 24\ (\frac{m^3}{d})} = 0.266\frac{kg}{m^3} = 266\ mg/l$$

– Xo: SS concentration at the biological basin inlet:

$$\text{Xo} = \frac{3{,}000{,}000\ (\text{person})\text{kx}\ 60\ (\frac{\frac{\text{g}}{\text{person}}}{\text{d}})}{6{,}940\ \text{x}\ 3.6\ \text{x}\ 24\ (\frac{\text{m}^3}{\text{d}})} = 300.2\ \text{mg/l}$$

– Xs: SS concentration at the treatment plant outlet = 20 $mg.L^{-1}$.

– b: endogenous respiration coefficient (0.07).

We then have:

$$\Delta Xb = 3.68\ [\ 0.6x\ 0.266 + 0.6\ x\ 0.300 - 0.020] - 0.07\ x\ 3.3$$

$$\Delta X_b = 0.945\ kg\ SM.m^3.d^{-1}$$

$$\Delta X_b/Cv = 0.945/0.99 = 0.954\ kg\ DS.kg^{-1}\ BOD_5$$

– Excess sludge: 0.954 x 160,000 = 152,640 kg $DS.d^{-1}$.

For a settled sludge concentration (Xb) of 8 kg/m^3, the flow rate of excess sludge to be extracted is obtained:

Excess sludge flow: 152,640/8 = 19,080 $m^3.d^{-1}$

Oxygenation

– Oxygen consumption:

$$OC = 0.5\ \rho\ Cv + 0.1\ X_b + Ch\ (4.6\ (N\text{-}NO_{3outlet}) + (4.6 - 2.9)\ N\text{-}NO_3denit)$$

with:

- ρ being the BOD_5 treatment efficiency;
- $N\text{-}NO_{3outlet}$ being the nitric nitrogen in treated water: 10 $mg.L^{-1}$ (estimation);
- $N\text{-}NO_3denit$ being the nitric nitrogen removal by denitrification: 8 $mg.L^{-1}$ (estimation):

$$OC = (0.5\ x\ 0.925\ x\ 0.99) + (0.1\ x\ 3.3) + 3.68\ (4.6\ x\ 0.010 + 1.7\ x\ 0.008)$$

$$OC = 1.01\ kg\ O_2.m^3.d^{-1}$$

– O_2 load under operating conditions:

$$\frac{\alpha OC}{Cv} = \frac{C_{O2}s}{C_{O2}s - C_{O2}b} OC / Cv$$

with:

- $C_{O2}s$ being the O_2 concentration at saturation (8.3 $mg.L^{-1}$ at 20°C);
- $C_{O2}b$ being the O_2 concentration in biological basins (2 $mgO_2.L^{-1}$).

Cv being the bolumetric loading (kg $BOD_5.m^{-3}.d^{-1}$)

$$\frac{\alpha OC}{Cv} = \frac{8.3}{8.3-2}\left(\frac{1.01}{0.99}\right) = 1.39 \frac{kgO_2}{kgBOD_5}$$

For the dimensioning of aeration devices, the O_2 loading is increased by a factor f = 1.40 to account for variations in concentrations and time:

$$\frac{\alpha OC}{Cv} = 1.39 \text{ x} 1.40 = 1.94 \frac{kgO_2}{kgBOD_5}$$

– Required oxygenation:

1.94. ($kgO_2.kg^{-1}$ BOD_5) x 160,000 (kg $BOD_5.d^{-1}$) = 310,400 kg $O_2.d^{-1}$

– Peak hourly oxygenation:

310,400/24 = 12,933 $kgO_2.h^{-1}$

Dimensioning of high-pressure air and stirring aeration equipment

– Air O_2 content: 280 $gO_2.m^{-3}$.

– Oxygenation efficiency: 3.5%/m.

– Insufflation depth: 4.5 m.

– Available O_2 content: 280 x 0.035 x 4.5 = 44.1 $gO_2.m^{-3}$.

– Peak air demand: 12,933 $kgO_2.h^{-1}$/0.0441 ($kgO_2.m^{-3}$) = 293,265 $m^3.h^{-1}$.

– Choosing eight blowers of 37,000 $m^3.h^{-1}$.

– O_2 transfer efficiency: OC/P = 1.7 $kgO_2.kWh^{-1}$.

– Choosing 64 high-pressure air and turbine agitator aeration units.

– Electrical power consumption per aeration unit:

$$\text{P electric absorbed} = \frac{12{,}933\ (\frac{\text{kgO}_2}{\text{h}})}{64 \text{ x } 1.7\ (\frac{\text{kgO}_2}{\text{kWh}})} = 119 \text{ kW}$$

– Energy requirements = 1.39 ($kgO_2.kg^{-1}$ BOD_5) x 160,000 (kg $BOD_5.d^{-1}$)/1.7 ($kgO_2.kWh^{-1}$).

– Energy requirements = 130,823 $kWh.d^{-1}$.

Clarification

Clarifiers are designed for wet weather flow rates:

2 Q24 = 2 x 6.94 = 13.88 $m^3.s^{-1}$ = 49,968 $m^3.h^{-1}$

– Activated sludge recirculation: 4.86 $m^3.s^{-1}$.

– Recirculation ratio: r = 4.86/13.88 = 0.35.

Since $r = \frac{Xb}{Xr - Xb}$, it is derived that:

$$Xb = \frac{r\,Xr}{1 + r} = \frac{0.35 \text{ x } 8}{1 + 0.35}$$

That is, X_b = 2.07 kg $DS.m^{-3}$.

– Comparative sludge volume:

$V_{sludge} = X_b$ x MI = 2.07 x 150 = 310.5 $ml.L^{-1}$

– Admissible surface loading:

Qs = 1.10 $m^3.m^2.h^{-1}$

– Clarifier area:

$$S = \frac{49{,}968\ (m^3/h)}{1.10\ (\frac{m}{h})} = 45{,}425\ m^2$$

– Choice: 60 circular tanks of 757 m^2 each, and 31.2 m diameter.

– Available area: 45,849 m^2.

– Clarifier depth: 4 m per unit.

– Total volume available: 45,849 x 4 = 183,396 m^3.

– Retention time Q_{24}:

$$ts = \frac{181{,}700(m^3)}{6{,}900 \text{ x } 3.6 \left(\frac{m^3}{h}\right)} = 7.3 \text{ h}$$

– Specific sludge loading $X_{bt} = X_b.qs$:

- for Q_{24}:

$$Xbt = \frac{3.3\left(\frac{kg}{m^3}\right) \text{ x } 6{,}900 \text{ x } 3.6\left(\frac{m^3}{s}\right)}{45{,}849\ (m^2)} = 1.78 \frac{kg\frac{MS}{m^2}}{h}$$

- for 2 Q_{24}:

$Xbt = 2.0 \text{ x } Qs = 2.0 \text{ x } 1.1 = 2.2 \text{ kg SM.m}^{-2}.h^{-1}$

6.9.4. *Example 4: population of 12,500 inhabitants*

Domestic wastewater

– Per capita consumption: 150 $L.d^{-1}$.

– Q_{24} = 12,500 x 0.150 = 1,875 $m^3.d^{-1}$ = 78.12 m^3/h = 0.0217 $m^3.s^{-1}$.

– Q max = 0.0217 x 1.33 = 0.0288 $m^3.s^{-1}$.

– Qd = 0.0217 x 1.18 = 0.0256 $m^3.s^{-1}$.

– Q night = 0.0217 x 0.87 = 0.0189 $m^3.s^{-1}$.

– Q wet weather = 2 Q max = 0.0434 $m^3.s^{-1}$.

Wastewater lifting station (Archimedes' screw)

– Peak flow rate = 0.0434 m^3/s = 43.4 $L.s^{-1}$.

– Elevation height: 7.50 m.

– Choice: two screws of 0.0217 $m^3.s^{-1}$. A backup screw of 0.0217 $m^3.s^{-1}$ will also be installed.

– Engine power:

$$PM = \frac{43.4\left(\frac{l}{s}\right) 7.50 \text{ m}}{0.65 \text{ x } 102} = 4.90 \text{ kW}$$

Screening – grit removal – oil removal

Aeration tank

– BOD_5 loading: the biological basins receive:

60 g BOD_5.person^{-1}.d^{-1}

– BOD_5: 12,500 person x 0.060 = 750 kg BDO_5.d^{-1}.

– Dimensioning:

- volumetric loading: Cv = 1.0 kg BOD_5.m^{-3}.d^{-1};
- sludge dry matter in tanks: Xb = 4.0 kg DS.m^{-3};
- mass loading: Cm = 0.30 kg BOD_5.kg^{-1} DS.d^{-1};
- Mohlman's index: MI = 150 mL.g^{-1};
- sludge volume: Vb = 500 mL.L^{-1}.

– Biological basin volume:

- total BOD_5/Cv = 750 (kg BOD_5.d^{-1})/1.0 (kg BOD_5.m^{-3}.d^{-1}) = 750 m^3;
- given two tanks: unit area 150 m², P = 5 m;
- Cv = 750 (kg BOD_5.d^{-1})/750 (m^3) = 1 (kg BOD_5.m^{-3}.d^{-1});
- hydraulic head: C_h = 0.0217 x 86,400/750 = 2.60 m^3.m^{-3}.d^{-1};
- aeration time, calculated from Q_{24}: 750 (m^3)/(0.0217 x 3.6 (m^3.h^{-1}) = 9.6 h;
- activated sludge recirculation ratio.

Concentration of activated sludge solids in recirculation:

$$Xr = \frac{1{,}200}{MI} = \frac{1{,}200}{150} = 8 \text{ kg DS/m}^3$$

Recirculation ratio:

$$Rr = \frac{Xb}{Xr - Xb} 100 = \frac{4.0}{8 - 4.0} 100 = 100\%$$

Recirculated activated sludge flow rate: 0.0217 x 1 = 0.0217 m^3.s^{-1}.

The recirculation pumps are designed for a recirculation ratio of 100%; that is, 0.0217 m^3.s^{-1}.

– Excess sludge production:

$$\Delta Xb = Ch\,[\,0.6\ Co + 0.6\ Xo - Xs] - b\ Xb$$

$a_m = 0.60$

– BOD_5o: BOD_5 concentration at the aeration tank inlet:

$$BOD_5o = \frac{750\ kgBOD_5/d}{0.0217\ x\ 3{,}600\ x\ 24\ (\frac{m^3}{d})} = 0.400\frac{kg}{m^3} = 400\ mg/l$$

– Xo: SS concentration solids at the biological basin inlet:

$$Xo = \frac{12{,}500\ (person)k\ x\ 60\ (\frac{\frac{g}{person}}{d})}{0.0217\ x\ 3{,}600\ x\ 24\ (\frac{m^3}{d})} = 400\ mg/l$$

– Xs: SS concentration at the treatment plant outlet = 20 $mg.L^{-1}$.

– b: endogenous respiration coefficient (0.07).

We then have:

$$\Delta X_b = 1.0\,[\,0.6\ x\ 0.400 + 0.6\ x\ 0.400 - 0.020] - 0.07\ x\ 4.0$$

$\Delta X_b = 0.18$ kg $DS.m^3.d^{-1}$

$\Delta X_b/Cv = 0.18/1 = 0.18$ kg $DS.kg^{-1}$ BOD_5

– Excess sludge: 0.18 x 750 = 135 kg $MS.d^{-1}$.

For a settled sludge concentration of 8 kg/m^3, we obtain the excess sludge flow rate to be extracted:

Excess sludge flow rate: 135/8 = 16.9 $m^3.d^{-1}$

Oxygenation

– Oxygen consumption:

$$OC = 0.5\ \rho\ Cv + 0.1\ X_b + Ch\ (4.6\ (N\text{-}NO_{3outlet}) + (4.6 - 2.9)\ N\text{-}NO_3denit)$$

with:

- ρ being the treatment efficiency in BOD_5;
- $N\text{-}NO_{3outlet}$ being the nitric nitrogen in treated water: 10 $mg.L^{-1}$ (estimation);

- $N\text{-}NO_3$denit being the nitric nitrogen removal by denitrification: 8 $mg.L^{-1}$ (estimation):

$$OC = (0.5 \times 0.925 \times 1) + (0.1 \times 4.0) + 1\,(4.6 \times 0.010 + 1.7 \times 0.008)$$

$$OC = 0.92 \text{ kg } O_2.m^{-3}.d^{-1}$$

– O_2 load under operating conditions:

$$\frac{\alpha OC}{Cv} = \frac{C_{O2}s}{C_{O2}s - C_{O2}b} OC / Cv$$

with:

- $C_{O2}s$ being the saturation O_2 concentration: 8.3 $mg.L^{-1}$ (at 20°C);
- $C_{O2}b$ being the O_2 concentration in aeration tanks: 2 $mgO_2.L^{-1}$.

$$\frac{\alpha OC}{Cv} = \frac{8.3}{8.3-2}\left(\frac{1.0}{1.0}\right) = 1.31 \frac{kgO_2}{kgBOD_5}$$

For the dimensioning of aeration devices, the O_2 loading is increased by a factor f = 1.40 to account for variations in concentrations and time:

$$\frac{\alpha OC}{Cv} = 1.31 \times 1.40 = 1.83 \frac{kgO_2}{kgBOD_5}$$

– Required oxygenation:

$$1.83\,(kgO_2.kg^{-1}\,BOD_5) \times 750\,(kg\,BOD_5.d^{-1}) = 1{,}372.5\text{ kg } O_2.d^{-1}$$

– Peak hourly oxygenation:

$$1{,}372.5/24 = 57.2\ kgO_2.h^{-1}$$

Dimensioning of high-pressure air and stirring aeration equipment

– Air O_2 content: 280 $gO_2.m^{-3}$.

– Oxygenation efficiency: 3.5%/m.

– Insufflation depth: 4.5 m.

– Available O_2 content: 280 x 0.035 x 4.5 = 44.1 g $O_2.m^{-3}$.

– Peak air demand: 57.2 $kgO_2.h^{-1}$/0.0441 ($kgO_2.m^{-3}$) = 2,522.5 $m^3.h^{-1}$.

– O_2 transfer efficiency: OC/P = 1.7 $kgO_2.kWh^{-1}$.

– Energy requirements = 1.82 (kg $O_2.kg^{-1}BOD_5$) x 750 (kg $BOD_5.d^{-1}$)/1.7 ($kgO_2.kWh^{-1}$).

– Energy requirements = 802 $kWh.d^{-1}$.

Clarification

Clarifiers are designed for wet weather flow rates:

2 Q24 = 2 x 0.0217 = 0.0434 $m^3.s^{-1}$ = 156.2 $m^3.h^{-1}$

– Activated sludge recirculation: 0.0217 $m^3.s^{-1}$.

– Recirculation ratio: Rr = 0.0217/0.0434 = 0.50.

Since Rr $= \frac{Xb}{Xr-Xb}$, it is deduced that:

$$Xb = \frac{Rr\ Xr}{1 + Rr} = \frac{0.50 \text{ x } 12}{1 + 0.50}$$

X_b = 4.0 kg $SM.m^{-3}$

– Comparative sludge volume:

$V_{sludge} = X_b \text{ x }_{MI} = 4 \text{ x } 150 = 600 \text{ ml.L}^{-1}$

– Admissible surface loading:

Qs = 1.10 $m^3.m^{-2}.h^{-1}$

– Clarifier area:

$$S = \frac{156.2(m^3/h)}{1.10\ (\frac{m}{h})} = 142\ m^2$$

– Clarifier depth: 4 m per unit.

– Total volume available: 142 x 4 = 568 m^3.

– Retention time Q_{24}:

$$t = \frac{568(m^3)}{21.7 \text{ x } 3.6\ (\frac{m^3}{h})} = 7.3\ h$$

– Specific sludge loading:

$$X_{bt} = X_b.Q$$

For Q_{24}, it is deduced that:

$$Xbt = \frac{4.0\left(\frac{kg}{m^3}\right) \text{ x } 21.7 \text{ x } 3.6 \left(\frac{m^3}{h}\right)}{142\ (m^2)} = 2.2\frac{kg\frac{MS}{m^2}}{h}$$

6.10. References

Alleman, J.L. and Irvine, R.L. (1980). Nitrification in sequencing batch biological reactor. *J. WPCF*, 52(11), 2747–2754.

American Society of Civil Engineers (ASCE) (2010). *Design of Municipal Wastewater Treatment Plants*, 5th edition. Manuals and Reports on Engineering Practice No. 76. Reston, Virginia.

Atlantic Canada (2006). *Atlantic Canada Wastewater Guidelines Manual for Collection, Treatment and Disposal*. ABL Environmental Consultants Ltd.

Béland, Y. (1984). Design d'un bioréacteur à boues actives basé sur les bilans de masse en matières volatiles en suspension (MVS) et en matières en suspension. *Sci. Techni. Eau*, 17(4), 21–32.

Benefield, L.D. and Randall, C.W. (1980). *Biological Process Design for Wastewater Treatment*. Prentice Hall, Englewood Cliffs.

Brosseau, C. (2015). Récupération de la matière organique biodégradable présente dans l'effluent d'un MBBR à forte charge. PhD Thesis, Polytechnique Montréal, Montreal.

Canler, J.-P. (2005). *Dysfonctionnement biologiques des stations d'épuration : origine et solutions*. Cemagref, Paris.

Deronzier, G. and Duchene, P. (1994). Influence of an horizontal flow on the performance of fine bubbles diffused air system. *Wat. Sci. Technol.*, 19(7), 239–247.

Deronzier, G. and Duchene, P. (1997). Vérification de la vitesse horizontale dans les chenaux d'aération pourvu d'un système d'aération par insufflation d'air. *Tech. Sci. Méthodes*, 3, 35–41.

Droste, R.L. and Gehr, R.L. (2019). *Theory and Practice of Water and Wastewater Treatment*. John Wiley & Sons, Hoboken, NJ.

Duchene, P. (1990). *Elimination de l'azote dans les stations d'épuration biologique des petites collectivités. Optimisation de la conception et du calcul des installations à boues activées en aération prolongée*. CEMAGREF, Antony.

Duchene, P. (1999). Estimation de la production de boues. Report, Cemagref.

Duchene, P. and Cotteux, E. (1998). La problématique des dysfonctionnements biologiques : bulking et moussage biologique en boues activées. *Tribune de l'eau*, 595, 78–96.

Duchene, P. and Cotteux, E. (2002). *Insufflation d'air fines bulles. Application aux stations d'épuration en boues activées des petites collectivités*. Cemagref, Paris.

Duchene, P. and Pujol, R. (1980). *Guide de lutte contre les mousses biologiques stables dans des eaux résiduaires. Théorie et technique*. CEBEDOC, Paris.

Duchene, P., Cotteux, E., Capela, S. (2000). Applying fine bubbles aeration to small aeration tanks. In *Proceedings 1st World Water Congress of the International Water Association (IWA)*, Paris, 409–415.

Eckenfelder, Jr. W.W. (1952). Aeration efficiency and design I. Measurement of oxygen transfer efficiency. *Sewage Ind. Wastes*, 24(10), 1221–1228.

Eckenfelder, Jr. W.W. (1959). Factors affecting the aeration efficiency of sewage and industrial wastes. *Sewage Ind. Wastes*, 31(1), 60–70.

Eckenfelder, Jr. W.W. (1981). *Gestion des eaux usées urbaines et industrielles*. Lavoisier, Paris.

Eckenfelder, Jr. W.W. (2006). *Wastewater Treatment*. Wiley, Boston.

Edeline, F. (1980). *L'épuration biologique des eaux résiduaires. Théorie et technologie*. Techniques et Documentation, Paris.

Ekama, G.A. and Wentzel, M.C. (2008). Organic material removal. In *Biological Wastewater Treatment: Principles Modelling and Design*, Chen, G., van Loosdrecht, M.C.M., Ekama, G.A., Brdjanovic, D. (eds). IWA Publishing, London.

Environmental Protection Agency (EPA) (1983). Design manual: Municipal wastewater stabilization ponds. Report No. EPA-25/1-83-015, Municipal Environmental Research Information, Cincinnati.

Felix, G.-O. and Emilio, G. (2009). Bioreactor scale-up and oxygen transfer rate in microbial processes: An overview. *Biotechnol. Adv.*, 27(2), 153–176.

Gerardi, M.H. and Lytle, B. (2010). *Troubleshooting the Sequencing Batch Reactor*. John Wiley & Sons, Hoboken.

Gillot, S. and Heduit, A. (2004). *Prédiction des capacités d'oxygénation en eau claire des systèmes d'insufflation d'air*. Cemagref, Paris.

Gomella, C. and Guerrée, H. (1982). *Les eaux usées dans les agglomérations urbaines ou rurales*. Eyrolles, Paris.

Grady, C.P., Daigger, G.T., Lim, H.C. (1999). *Biological Wastewater Treatment*. Marcel Dekker Inc., New York.

Grady, C.P., Daigger, G.T., Love, N.G., Filipe, C.D.M. (2011). *Biological Wastewater Treatment*. IWA Publishing/CRC Press, Boca Raton.

Helness, H., Melin, E., Ulgenes, Y., Jarvinen, P., Rasmussen V., Ødegaard, H. (2005). High-rate wastewater treatment combining a moving bed biofilm reactor and enhanced particle separation. *Water Sci. Technol.*, 52(10/11), 117–127.

Irvine, R.L. (1977). Application of sequencing batch reactors for treatment of municipal and industrial wastewater. Report, NSF/RA-770259, Department of Civil Engineering, University of Notre-Dame, Notre-Dame.

Irvine, R.L. and Busch, A. (1979). Sequencing batch biological reactor: An overview. *J. WPCF*, 51(2), 235–244.

Irvine, R.L., Wilderer, P.A., Flemming, H.C. (1997). Controlled unsteady state processes and technologies: An overview. *Water Sci. Technol.*, 35(1), 1–10.

Ivanovic, I. and Leiknes, T.O. (2012). Particle separation in moving bed biofilm reactor – Applications and opportunities. *Separation Sci. Technol.*, 47(5), 647–653.

Jenkins, D. (1980). Discussion on bulking deflocculation and pin point floc. *J. WPCF*, 52(3), 622–624.

Jenkins, D. (1986). *Manual of Causes and Control of Activated Sludge Bulking and Foaming.* Ridgeline Press, Lafayette.

Jenkins, D., Richards, M.G., Daigger, G.T. (2003). *The Causes and Cures of Activated Sludge Bulking and Foaming.* Lewis Publishers, Ann Arbor.

Jenkins, D., Richards, M.G., Daigger, G.T. (2004). *Manual on the Causes and Control of Activated Sludge Bulking, Foaming, and Other Solids Separation Problems.* Lewis Publishers, Boca Raton.

Ketchum, Jr. L.H. (1997). Design and physical features of sequencing batch reactor. *Water Sci. Technol.*, 35(1), 11–18.

Kossy, K. and Al-Ahmady, K. (2006). Analysis of oxygen transfer performance on sub-surface aeration systems. *Int. J. Environ. Res. Public Health*, 21–36.

Krause, P.E. (2010). *Design of Municipal Wastewater Treatment Plants, Volume 2: Liquid Treatment Processes.* Manual of Practice No. 8. WEF Press, Alexandria.

Manning, J.F. and Irvine, I.R. (1985). The biological removal of phosphorus in a sequencing batch reactor. *J. WPCF*, 57(1), 87–94.

Metcalf & Eddy, Inc. (2003). *Wastewater Engineering: Treatment and Reuse.* McGraw-Hill, New York.

Metcalf & Eddy, Inc. and Aecom (2007). *Water Reuse: Issues, Technologies, and Applications.* McGraw-Hill, New York.

Metcalf & Eddy, Inc. and Aecom (2014). *Wastewater Engineering Treatment and Resource Recovery.* McGraw-Hill, New York.

Ministère de l'environnement et du cadre de vie (1979). Techniques et économie de l'épuration des eaux résiduaires. Ministère de l'environnement et du cadre de vie, Paris.

Modin, O., Saheb Alam, D., Persson, F., Wilen, B.M. (2015). Sorption and release of organics by primary, anaerobic, and aerobic activated sludge mixed with raw municipal wastewater. *Plos One*, 10(3), 1–15.

Monod, J. (1941). *Recherches sur la croissance des cultures bactériennes*. Hermann, Paris.

Mueller, J.A., Boyle, W.C., Pöpel, H.J. (2002). *Aeration: Principles and Practices*. CRC Press, New York.

Ontario Ministry of the Environment (MOE) (2008). *Design Guidelines for Sewage Works*. Ontario Ministry of the Environment, Canada.

Pellas, L., Beutler, E., Gilles, P. (1995). Modélisation déterministe des boues activées et méthodes de dimensionnement classiques. *Tech. Sci. Méthodes*, 9, 679–682.

Poltak, R.F., Jennings, M., Murphy, J., Dupuis, R.J. (2005). Sequencing batch reactor design and operational considerations. New England Interstate Water Pollution Control Commission, Lowell.

Pujol, R., Duchene, P., Canler, J.-P. (1991). Biological foaming activated sludge plants: Characterization and situation. *Water Res.*, 25(11), 1399–1404.

Qasim, S.R. and Zhu, G. (2018a). *Wastewater Treatment and Reuse: Theory and Design Examples, Volume 1. Principles and Basic Treatment*. CRC Press, Boca Raton.

Qasim, S.R. and Zhu, G. (2018b). *Wastewater Treatment and Reuse: Theory and Design Examples, Volume 2. Post-Treatment, Reuse, and Disposal*. CRC Press, Boca Raton.

Roberge, F. (1991). Analyse de l'influence des conditions d'opération d'un réacteur biologique séquentiel sur le biotraitement d'une eau usée municipale faiblement chargée, mémoire de maîtrise. PhD Thesis, École Polytechnique de Montréal, Montréal.

Sahtout, N. (2012). Etude de station de traitement des eaux à boues activées : optimisation de l'aération et la clarification en vue d'une bonne décantabilité des boues. PhD Thesis, Badji Mokhtar Annaba University, Annaba.

Satin, M. and Selmi, B. (2006). *Guide technique de l'assainissement*. Le Moniteur, Paris.

Stentrom, M., Leu, S., Jlang, P. (2006). Theory to practice: Oxygen transfer and the new ASCE standard. *Proc. Water Environ. Fed.*, 4838–4852.

Tchobanoglous, G., Burton, F.L., Stensel, H.D. (2003). *Wastewater Engineering: Treatment and Reuse*, 4th edition. McGraw-Hill, Boston.

Thomazeau, R. (1981). *Stations d'épuration : eaux potables, eaux usées*. Technique & Documentation, Paris.

Van Winckel, T. (2014). Development of high-rate activated sludge processes for energy-efficient wastewater treatment. PhD Thesis, University of Ghent, Gent.

Von Sperling, M. (2007). *Activated Sludge and Aerobic Biofilm Reactors: 5*. IWA Publishing, London.

Water Environment Federation (2006). *Membrane Systems for Wastewater Treatment.* WEF Press/McGraw-Hill, New York.

Water Environment Federation (2007). *Operation of Municipal Wastewater Treatment Plants.* WEF Manual of Practice No. 11. McGraw-Hill, New York.

Water Environment Federation and American Society of Civil Engineers/Environmental and Water Resources Institute (2005). *Biological Nutrient Removal (BNR) Operation in Wastewater Plants.* Manual of Practice No. 29. McGraw-Hill, New York.

Water Environment Federation and American Society of Civil Engineers/Environmental and Water Resources Institute (2010). *Design of Municipal Wastewater Treatment Plants.* WEF Manual of Practice No. 8. McGraw-Hill, New York.

Water Environment Federation and American Society of Civil Engineers/Environmental and Water Resources Institute (2018). *Design of Water Resource Recovery Facilities.* Manual of Practice No. 8. McGraw-Hill, New York.

Zhen, H., Anurak, P., Warawitya, M. (2003). Oxygen transfer measurement in clean water. *J. KMITNB*, 13(1), 18–29.

Index

A, B

activated sludge, 33, 45, 47–49, 51–53, 125, 179, 181–184, 190, 192, 194–196, 199, 200, 203–206, 210, 218, 225–227, 230–233, 249, 251, 254, 269, 270, 272, 275, 278, 279, 284, 286, 287, 290, 291, 293, 298, 301, 303, 306
aeration, 24, 53, 59, 92, 93, 102, 183, 184, 186, 187, 190, 191, 193–198, 202, 205, 206, 215, 216, 222, 224, 228–230, 234, 235, 242, 243, 245, 246, 248–250, 253–261, 265, 266, 268, 270–273, 275–280, 284–287, 289–294, 297, 298, 300, 303–305
  permanent, 255, 259–261, 287
  sequenced, 255, 259, 261
air diffusers, 93, 249, 279, 286
ammonia, 15–17, 45, 61
bacteria, 1, 2, 6, 7, 25, 26, 29–32, 34–36, 44, 45, 47, 52, 53, 55–57, 180, 181, 183, 185, 216, 223, 225, 228–233, 270, 277, 280, 281, 284, 285, 289
Bioring™, 289–291
brushes, 82, 249–251, 254, 255, 280

C, D

coliforms, 26, 29, 30, 34–36
  fecal, 29, 34–37
  total, 34–36
comminution, 79, 89, 90
complete mix, 158, 192, 194, 199, 200, 213, 279, 281, 286
demand
  biological oxygen (BOD, BOD_5), 1, 6, 8
  chemical oxygen (COD), 1, 6, 10, 57, 126
design/dimensioning, 24, 67, 107, 108, 111, 114, 133–135, 139, 146–148, 153, 186, 192, 193, 221, 222, 227, 259, 278, 291, 300, 305
drugs, 46, 58

E, F, G

endocrine disruptors, 2, 20, 23, 46
enterococci, 35, 37
flow piston/plug-flow, 113, 152, 280, 281
gradual feeding, 283
grease
  and oil trap, 101
  trap, 95

H, I, M

helminths, 1, 26, 29, 30, 32–34, 36, 46

hormones, 21
integral mix, 194
microorganisms, 1, 2, 5, 6, 9, 10, 15, 19, 26, 28, 29, 32–34, 36–38, 44, 46, 53–55, 57, 60, 76, 179, 181–184, 226, 233, 270, 278, 280, 281, 284

N, O

nitrates, 7, 15–17, 45, 46, 55, 180, 228
nitrogen, 7, 11, 15–17, 24, 44, 48, 51, 53–56, 60, 61, 75, 179, 193, 196, 222, 223, 228, 230, 232, 233, 277, 280, 283, 290, 299, 305
oxidation channel, 213, 279, 281
oxygen
 requirements, 255, 256, 292, 294
 transfer, 234, 236, 242, 243, 249, 251, 256, 266, 270

P, S

pesticides, 2, 5, 20, 21, 46, 58, 60
phosphates, 9, 19, 45, 55, 61, 126, 180
phosphorous, 4, 14, 19, 20, 24, 44, 45, 48, 51, 54–56, 126, 168, 179, 223, 230, 280, 290
pollutant loads, 44, 65, 107, 143
pretreatment, 48, 51, 62, 91, 280
principle, 6, 14, 80, 93, 122, 129, 134, 155, 179, 218
protozoa, 1, 26, 29, 31, 32, 34–37, 232
protozoans, 38
sand trap, 91, 93, 96, 97, 99, 100
 aerated, 100
screen(s), 80, 84–89
 curved, 81
 mechanical, 81, 89
 straight, 81
settler
 circular, 109, 118, 124, 125
 rectangular, 123, 125
settling, 45, 52, 54, 56, 58, 96, 107, 108, 111, 113, 116, 120–122, 125–129, 131, 133–135, 137, 138, 141, 144–146, 148–152, 154–157, 160, 161, 164–168, 171, 179, 182, 194, 206, 208, 210, 222, 224, 227, 228, 232, 278, 280, 284, 285, 287–289
 lamellar, 52, 134, 138, 145, 154, 155
 primary, 45, 52, 107, 120, 125, 128, 138, 157, 168, 171, 194, 280, 284
 velocities, 139, 156, 160
sieving, 51, 80, 90
suspended solids, 1, 3–5, 24, 43, 45, 48, 52, 59, 76, 90, 107, 108, 112, 115, 120, 125, 126, 129, 134, 138, 144, 149, 160–163, 167, 168, 193, 194, 196, 199, 202, 204, 222, 223, 227, 236, 240, 246, 268–270, 284, 290
 volatile, 4, 5, 24, 183, 195, 196, 215

T, U, V

total organic carbon, 6, 13
treatment
 primary, 48, 52, 129, 146, 157
 secondary, 48, 52
turbines, 55, 144, 249–251, 255, 261, 279, 280, 286
unitary operation, 126
viruses, 1, 2, 26–28, 30–32, 34–37

Other titles from

ISTE

in

Ecological Science

2025

COURANT Dimitri, REBER Bernard
Deliberative Democracy and Ecological Transition: The French Citizens' Convention for Climate
(Planetary Changes, Decision, Governance Set – Volume 1)

2024

AMIARD Jean-Claude, ZERBIB Jean-Claude
Radioactive Risk for Humans
(Radioactive Risk Set – Volume 7)

AKBAR Saima, KUMAR SINGH Suraj, KANGA Shruti
Disaster Management: Preparedness and Mitigation

FLEURENCE Joël
Aquatic Biotechnologies: From Genetic Engineering to Enzymatic or Fermentation Engineering

GAID Kader
Wastewater Reuse 1: Characteristics, Uses, Applications, Filtration and Disinfection of Water

GAID Kader
Wastewater Reuse 2: Micropollutants, Membranes and Treatment Procedures

MATAGNE Patrick
Birth of Scientific Ecology: Eugenius Warming (1841–1924)

2023

GAID Kader
Drinking Water Treatment 1: Water Quality and Clarification
Drinking Water Treatment 2: Chemical and Physical Elimination of Organic Substances and Particles
Drinking Water Treatment 3: Organic and Mineral Micropolluants
Drinking Water Treatment 4: Membranes Applied to Drinking Water and Desalination
Drinking Water Treatment 5: Calco-carbonic Equilibrium and Disinfection

2022

AMIARD Jean-Claude
Marine Radioecology
(Radioactive Risk Set – Volume 6)

GODET Laurent, DUFOUR Simon, ROLLET Anne-Julia
The Baseline Concept in Biodiversity Conservation: Being Nostalgic or Not in the Anthropocene Era

GRISON Claude, CASES Lucie, LE MOIGNE Mailys, HOSSAERT-MCKEY Martine
Photovoltaism, Agriculture and Ecology: From Agrivoltaism to Ecovoltaism

LÉVÊQUE Christian
Biodiversity Erosion: Issues and Questions

2021

AMIARD Jean-Claude
Disarmament and Decommissioning in the Nuclear Domain
(Radioactive Risk Set – Volume 4)
Management of Radioactive Waste
(Radioactive Risk Set – Volume 5)

FLEURENCE Joël
Microalgae: From Future Food to Cellular Factory

JOLY Fernand, BOURRIÉ Guilhem
Mankind and Deserts 1: Deserts, Aridity, Exploration and Conquests
Mankind and Deserts 2: Water and Salts
Mankind and Deserts 3: Wind in Deserts and Civilizations

PIOCH Sylvain, SOUCHE Jean-Claude
Eco-design of Marine Infrastructures: Towards Ecologically-informed Coastal and Ocean Development

2020

BRUSLÉ Jacques, QUIGNARD Jean-Pierre
Fish Behavior 1: Eco-ethology
Fish Behavior 2: Ethophysiology

ROSSIGNOL Jean-Yves
Climatic Impact of Activities: Methodological Guide for Analysis and Action

2019

AMIARD Jean-Claude
Industrial and Medical Nuclear Accidents: Environmental, Ecological, Health and Socio-economic Consequences
(Radioactive Risk Set – Volume 2)
Nuclear Accidents: Prevention and Management of an Accidental Crisis
(Radioactive Risk Set – Volume 3)

BOULEAU Gabrielle
Politicization of Ecological Issues: From Environmental Forms to Environmental Motives

GIRAULT Yves
UNESCO Global Geoparks: Tension Between Territorial Development and Heritage Enhancement

KARA Mohamed Hichem, QUIGNARD Jean-Pierre
Fishes in Lagoons and Estuaries in the Mediterranean 2: Sedentary Fish
Fishes in Lagoons and Estuaries in the Mediterranean 3A: Migratory Fish
Fishes in Lagoons and Estuaries in the Mediterranean 3B: Migratory Fish

OUVRARD Benjamin, STENGER Anne
Incentives and Environmental Policies: From Theory to Empirical Novelties

2018

AMIARD Jean-Claude
Military Nuclear Accidents: Environmental, Ecological, Health and Socio-economic Consequences
(Radioactive Risk Set – Volume 1)

FLIPO Fabrice
The Coming Authoritarian Ecology

GUILLOUX Bleuenn
Marine Genetic Resources, R&D and the Law 1: Complex Objects of Use

KARA Mohamed Hichem, QUIGNARD Jean-Pierre
Fishes in Lagoons and Estuaries in the Mediterranean 1: Diversity, Bioecology and Exploitation

2016

BAGNÈRES Anne-Geneviève, HOSSAERT-MCKEY Martine
Chemical Ecology

2014

DE LARMINAT Philippe
Climate Change: Identification and Projections

Printed and bound by CPI Group (UK) Ltd, Croydon, CR0 4YY

26/08/2026

14949808-0002